ADVANCES IN VIRAL ONCOLOGY

Volume 7

Advances in Viral Oncology
Volume 7

Analysis of Multistep Scenarios in the
Natural History of Human or Animal Cancer

Editor

George Klein, M.D., D.Sc.

Department of Tumor Biology
Karolinska Institutet
Stockholm, Sweden

Raven Press ❦ New York

Raven Press, 1185 Avenue of the Americas, New York, New York 10036

© 1987 by Raven Press Books, Ltd. All rights reserved. This book is protected by copyright. No part of it may be reproduced, stored in a retrieval system, or transmitted, in any form or by any means, electronic, mechanical, photocopying, recording, or otherwise, without the prior written permission of the publisher.

Made in the United States of America
987654321

Library of Congress Cataloging-in-Publication Data
Main entry under title:

Analysis of multistep scenarios in the natural history of human or animal cancer.

   (Advances in viral oncology; v. 7)
   Includes bibliographies and index.
   1. Carcinogenesis. 2. Oncogenes. 3. Cancer—
Genetic aspects. 4. Viral carcinogenesis.
I. Klein, George, 1925–   . II. Series.
[DNLM: 1. Cell Transformation, Neoplastic.
2. Neoplasms–etiology. W1 AD888 v.7 / QZ 202 M961]
RC268.5.M85   1986   616.99'4071   86-26032
ISBN 0-88167-191-6

The material contained in this volume was submitted as previously unpublished material, except in the instances in which credit has been given to the source from which some of the illustrative material was derived.

Great care has been taken to maintain the accuracy of the information contained in the volume. However, Raven Press cannot be held responsible for errors or for any consequences arising from the use of the information contained herein.

Materials appearing in this book prepared by individuals as part of their official duties as U.S. Government employees are not covered by the above-mentioned copyright.

Preface

Volume 6 of this series dealt with the experimental analysis of multifactorial interactions in the tumorigenic process. This volume focuses on the corresponding experiments of nature, the spontaneous multistep development of human and animal cancers. We have selected some scenarios where one or several steps have been defined in terms of specific genes and/or phenotypic traits.

The ingenious prediction of Alfred G. Knudson, postulating that both alleles of the same gene are damaged or lost in some childhood tumors with a strong hereditary component has been fully confirmed by modern molecular analysis for retinoblastoma and Wilms' tumor. This is so far the most definitive application of molecular biology to the analysis of tumor development in any system. It also demonstrates that the same genetic change, the loss of both alleles of a critically important (rbl-1, retinoblastoma) gene, can arise by a variety of cytogenetic mechanisms, such as deletion, chromosome loss with or without the duplication of the remaining homologue, and somatic crossing over. In fact, every conceivable mechanism that could account for the postulated gene loss and was testable by the molecular techniques, RFLP marker analysis in particular, was actually found to occur in one tumor or another.

The genes that must become nonfunctional in the highly specialized precursor cells of retinoblastoma or Wilms' tumor to trigger malignant development are sometimes referred to as antioncogenes, since the presence of the normal allele prevents tumor development at the level of the somatic target cell. Their mechanism of action is probably quite different from the tumor suppressor genes revealed by reversion or somatic hybridization experiments, however. According to Knudson's original hypothesis, they may control a terminal differentiation step that brings proliferating cells to a final halt, with irreversible loss of mitotic activity. In line with the concept that tumor progression is never final, it is interesting to note that progression is often accompanied by the amplification of another gene, N-*myc,* in retinoblastoma and neuroblastoma as well.

A second area, oncogene activation by chromosomal translocation, has been particularly relevant for the genesis of certain B-lymphocyte-derived tumors in three different species: humans, mice, and rats. The homologies between the Ig/*myc* juxtapositions of these tumors are remarkable in spite of vast differences in inducing agent, natural history, site of predilection, and histological subtype of the ultimate B-cell tumor. The regular occurrence of these translocations is the best evidence that they represent an essential, i.e., rate-limiting step, in the tumorigenic process. No qualitative structural change of the *myc*-coding exons appears to be required. Constitutive activation of the oncogene by deregulation is probably both necessary and sufficient. Immortalizing viruses, such as Epstein-Barr virus in humans and Abelson virus in mice, may shorten the latency period and increase the frequency of the final tumor. These viruses probably act by expanding the target cell population at risk, rather than directly by promoting the translocation event.

The other major known tumor-associated translocation is reflected by the oldest known nonrandom anomaly in any tumor: the Ph1 chromosome. It acts through the production of an abnormal fusion protein, contributed by sequences on both sides of the breakpoint in the typical (9;22) translocation. The variant translocations that appear to arise by a reciprocal exchange between chromosome 22 and an autosome other than 9, turned out to arise by three-way exchanges, with regular transfer of the c-*abl* oncogene, localized at the tip of chromosome 9, to chromosome 22. The result is thus the same; the production of an abnormal 210 K fusion protein. Unlike the translocation-deregulated *myc* system where the normal allele is turned off, the 210 K c-*abl*/*bcr* fusion protein is expressed together with its normal counterpart. It is intriguing that this very limited material of two major systems has already revealed that translocation-activated oncogenes can work through either one of the two alternative models known from the experimental systems, deregulation of a normal gene, or structural changes that lead to the production of an abnormal protein.

The mouse mammary tumor system represents the oldest and best analyzed multistep scenario in the history of cancer research, where viral and cellular changes, some of them triggered by hormonal signals, contribute to tumor development by increasing the probability of the ultimate step toward autonomous growth.

Oncogene amplification is usually found during a late stage of tumor progression. Its significance for the appearance of a more metastatic and invasive phenotype has been particularly well documented for small-cell lung carcinomas (SCLC). Phenotypically still undefined amplifications have also been found in a number of other tumors. The *myc* oncogene family is particularly conspicuous among the amplifications so far reported (at least 90% of all positive reports). Amplified *myc* genes were found in tumors of all major tissues, hemopoietic, epithelial, mesenchymal, and neuronal. The SCLC scenario may lead the way toward an understanding of the selective value of the *myc* amplifications for the tumor cell, partly because of its association with a more metastatic variant cell type and because the three major members of the *myc* family, c-, N- and L-*myc* can amplify alternatively.

The multistep scenarios described thus far show either sequential activation of different oncogenes or the loss of an antioncogene, followed by the amplification of an oncogene or composite interactions between viral transformation and cellular oncogene activation. Mutations, deletions, translocations, and gene amplifications have all been shown to participate. This is understandable, since selection must act at the level of the cellular phenotype, not at the level of the cytogenetic mechanism. Although this may appear complex at first, there is no doubt that it is merely a very modest beginning in relation to the biological reality that will eventually become unraveled. It is to be expected that all three gene worlds that now can be discerned by their outlines, oncogenes, tumor suppressor genes, and modulators of neoplastic behavior will take part in this complex microevolution.

This volume will be of interest to all research oncologists, as well as to virologists, immunologists, and cellular and molecular biologists.

GEORGE KLEIN

Contents

1 **A Two-Mutation Model for Human Cancer**
Alfred G. Knudson, Jr.

19 **Recessive Human Cancer Susceptibility Genes (Retinoblastoma and Wilms' Loci)**
William F. Benedict

35 **Molecular Basis of Human B- and T-Cell Neoplasia**
Carlo M. Croce, Jan Erikson, Yoshihide Tsujimoto, and Peter C. Nowell

53 **Multistep Cytogenetic Scenario in Chronic Myeloid Leukemia**
Sverre Heim and Felix Mitelman

77 **Activation of c-*abl* as a Result of the Ph' Translocation in Chronic Myelocytic Leukemia**
J. Groffen, N. Heisterkamp, and K. Stam

99 **Plasmacytoma Development in BALB/c Mice**
Michael Potter

123 **Multistep Model of Mouse Mammary Tumor Development**
David W. Morris and Robert D. Cardiff

141 **Role of Oncogene Amplification in Tumor Progression**
Yoichi Taya, Masaaki Terada, and Takashi Sugimura

155 **Amplification and Expression of the *myc* Gene in Small-Cell Lung Cancer**
Burke J. Brooks, Jr., James Battey, Marion M. Nau, Adi F. Gazdar, and John D. Minna

173 **Burkitt's Lymphoma, A Human Cancer Model for the Study of the Multistep Development of Cancer: Proposal for a New Scenario**
Gilbert M. Lenoir and Georg W. Bornkamm

207 **In Defense of the "Old" Burkitt Lymphoma Scenario**
George Klein

213 **Subject Index**

Contributors

James Battey
NCI-Navy Medical Oncology Branch
National Cancer Institute
National Institutes of Health & Naval
 Hospital and the Department of
 Medicine
Uniformed Service University of the Health
 Sciences
Bethesda, Maryland 20814

William F. Benedict
Division of Hematology-Oncology and
 The Clayton Ocular Oncology Center
Children's Hospital of Los Angeles
Los Angeles, California 90027

Georg W. Bornkamm
Institut für Virologie im
 Zentrum für Hygiene
7800 Freiburg—i.B.,
Federal Republic of Germany

Burke J. Brooks, Jr.
4700 Chastaut Street
Metairie, Louisiana 70006

Robert D. Cardiff
Department of Pathology
University of California
Davis, California 95616

Carlo M. Croce
The Wistar Institute
3601 Spruce Street
Philadelphia, Pennsylvania 19104

Jan Erikson
The Wistar Institute
3601 Spruce Street
Philadelphia, Pennsylvania 19104

Adi F. Gazdar
NCI-Navy Medical Oncology Branch
National Cancer Institute
National Institutes of Health & Naval
 Hospital and the Department of
 Medicine
Uniformed Service University of the Health
 Sciences
Bethesda, Maryland 20814

J. Groffen
Laboratory of Molecular Genetics
Oncogene Science, Inc.
222 Station Plaza North
Mineola, New York 11501

Sverre Heim
Department of Clinical Genetics
University Hospital
S-221 85 Lund
Sweden

N. Heisterkamp
Laboratory of Molecular Genetics
Oncogene Science, Inc.
222 Station Plaza North
Mineola, New York 11501

George Klein
Department of Tumor Biology
Karolinska Institute
Box 60400
104 01 Stockholm, Sweden

Alfred G. Knudson, Jr.
Institute for Cancer Research
Fox Chase Cancer Center
Philadelphia, Pennsylvania 19111

Gilbert M. Lenoir
International Agency for Research on
 Cancer
69372 Lyon Cedex 08, France

John D. Minna
NCI-Navy Medical Oncology Branch
National Cancer Institute

National Institutes of Health & Naval
  Hospital and the Department of
  Medicine
Uniformed Service University of the Health
  Sciences
Bethesda, Maryland 20814

Felix Mitelman
Department of Clinical Genetics
University Hospital
S-221 85 Lund
Sweden

David W. Morris
Department of Pathology
University of California
Davis, California 95616

Marion M. Nau
NCI-Navy Medical Oncology Branch
National Cancer Institute
National Institutes of Health & Naval
  Hospital and the Department of
  Medicine
Uniformed Service University of the Health
  Sciences
Bethesda, Maryland 20814

Peter C. Nowell
Department of Pathology and Laboratory
  Medicine
University of Pennsylvania School of
  Medicine
Philadelphia, Pennsylvania 19104

Michael Potter
National Cancer Institute
National Institutes of Health
Bethesda, Maryland 20892

K. Stam
Laboratory of Molecular Genetics
Oncogene Science, Inc.
222 Station Plaza North
Mineola, New York 11501

Takashi Sugimura
National Cancer Center Research Institute
Tsukiji, Chuo-ku
Tokyo 104, Japan

Yoichi Taya
National Cancer Center Research Institute
Tsukiji, Chuo-ku
Tokyo 104, Japan

Masaaki Terada
National Cancer Center Research Institute
Tsukiji, Chuo-ku
Tokyo 104, Japan

Yoshihide Tsujimoto
The Wistar Institute
3601 Spruce Street
Philadelphia, Pennsylvania 19104

ADVANCES IN VIRAL ONCOLOGY

Volume 7

A Two-Mutation Model For Human Cancer

Alfred G. Knudson, Jr.

*Institute for Cancer Research, Fox Chase Cancer Center,
Philadelphia, Pennsylvania 19111*

GENERAL CONSIDERATIONS

Mutation and Cancer

The causation of cancer by agents known also to be mutagenic has led most investigators to conclude that at least one etiologic event is a somatic mutation. This conclusion has in turn focused attention on the nature of the genes that undergo such mutations and on the number of events that may be genetic. Information bearing on these questions has come from the study of chromosomal abnormalities in cancers, of oncogenic viruses, and of hereditary cancers. Two broad classes of genes have been discovered so far, the first being the oncogenes, revealed by oncogenic viruses, and the second, a different class, revealed by the study of hereditary cancer in children. The purpose of this review is to show how the incidences of certain cancers may result from two genetic events, how the hereditary and nonhereditary forms relate to each other, and how such tumors can be attributed to mutations in a new class of cancer genes.

Age-Specific Incidences

Early stochastic models that related age-specific incidences of the common cancers of adults to cellular events generally concluded that five to seven such events were involved. Armitage and Doll (1), who proposed one such model, also proposed a model in which cells that had sustained an initial event might grow exponentially, and they concluded that two events could account for the observed incidences (2). No model that proposes one event would suffice for the common cancers unless the underlying process or the number of target cells was steadily changing with time; there is no experimental evidence in support of either of these possibilities. So there was a discrepancy among the models, the estimated number of events ranging from two to seven. Furthermore, none of the models took into account the origin and renewal of tissue target cells.

All of the above models were inadequate for a description of cancers with peak incidence in childhood or of cancers, such as those of breast or ovary, that show a decrease in the slope of incidence during adult life. Burch (12) proposed a model that took into account the growth of normal tissues, as well as differential growth of cells in intermediate stages. Knudson et al. (36,47) considered cellular multiplication and differentiation in a model for the origin of embryonal tumors, such as retinoblastoma. It was clear that the kinetics of normal target cells was the principal determinant of the incidence curves for embryonal tumors.

Taking into consideration all of these variables—kinetics of target cells, rates of determining events, and behavior of intermediate cells—Moolgavkar and his colleagues (65,66) developed a generalized two-event model that is consistent with incidence data on common cancers, variant cancers such as that of the breast, and embryonal cancers of childhood.

Hereditary Cancers

That cancer can be heritable has been known since the previous century, and at least 50 examples are now known (68), some of the best known being polyposis coli (with carcinoma of the colon), neurofibromatosis (with various tumors), hereditary retinoblastoma, and xeroderma pigmentosum (XP) (with skin cancers, melanomas, and some internal cancers).

In some instances, predisposition to cancer is recessively inherited. Persons with XP are extremely susceptible to cancers inducible by sunlight, although the apparent increase in internal cancers suggests that sunlight is not an exclusive agent in the disease (55). Protection from sunlight can prevent skin cancer (60). Cells *in vitro* are sensitive to ultraviolet light and have been reported to show an increased specific-locus mutation rate (62). There is defective repair of damage to DNA (15). In terms of any mutational model for cancer, XP may be considered as a condition that greatly increases the rate at which one or more of the carcinogenic events occurs.

Another recessively inherited condition that predisposes to cancer is Bloom's syndrome (BS). Increased spontaneous mutation rates have been reported for BS (85), which could account for the predisposition to several kinds of cancer. In addition, cells from patients show a very high incidence of sister chromatid exchanges, and peripheral blood cells show homologous chromosomal exchange (14,31). The latter phenomenon could convert a cell that is heterozygous for a somatic mutation in a cancer gene into a homozygous cell, thereby permitting expression of a recessive "cancer gene" (24,43,71).

Neither XP or BS involves mutation of a locus that is one of the "events" on the path to cancer. On the other hand, the dominantly inherited predispositions to cancer have been considered to be just this (42,44,52), and it is these genetic conditions that are discussed in connection with a two-event model of cancer.

A TWO-EVENT MODEL OF CANCER

The Target Cell

None of the models proposed for adult cancer was satisfactory for the embryonal tumors of children, such as retinoblastoma, Wilms' tumor, and neuroblastoma. For each of these tumors, the age-specific incidence peaks in early life, then declines, and falls to such low levels that these tumors are never, or only very rarely, found in adults. The explanation for this phenomenon evidently lies in the observation that these tumors arise from cell types that are normally present in embryonic, fetal, or early postnatal tissues, but which characteristically differentiate into postmitotic cells. Thus, the retinoblast is a precursor cell for neural and photoreceptor cells in the retina; the nephroblast, for the nephrons; and the neuroblast, for the adrenal medullae and ganglia of the automic nervous system. The oncogenic event(s) must occur before the postmitotic state is reached. This conclusion leads directly to a necessary consideration of the kinetics of the target cell population in any model.

Although the kinetics of these precursor cells is not known in detail, useful observations have been made. Consider the development of the retina, for example. The first problem is the number of cells committed to form a retina. Using mice that were allophenic for normal retina and hereditary degeneration of the retina, Mintz and Sanyal (64) found sectors of degenerate retina. Since the smallest sector was one-tenth of the total retina, and other sectors seemed to be integral multiples thereof, the investigators concluded that 10 cells are committed to being retinoblasts and therefore precursors of the whole retina. This number is not known for man. The final number of cells derived from these retinoblasts is of the order of magnitude of 10^8 in man, the single largest fraction being the photoreceptor cells. Most of these cells are formed prenatally, since the retina is well developed at birth; however, some cells are still proliferating at birth, and it is not known precisely when these divisions cease. If the initial number of committed retinoblasts in the human eye is also of the order of magnitude 10, then approximately 10^7 cell divisions occur among retinoblasts in each eye during development.

The distribution of retinoblasts as a function of time can be surmised in its general form. No retinoblasts are present at conception and none are present again at some later (unknown) age in childhood; however they must certainly be present before 3 years of age. The peak age for cells still at the retinoblast stage is before birth, probably some time during the last trimester of pregnancy. Several mathematical functions, including the gamma density function, satisfy these general requirements.

Not as much is known quantitatively about the precursor cells for Wilms' tumor and neuroblastoma, but it is known that their generation reaches a peak during early life, since such cells are not found in older children or in adults.

A First Mutational Event

Hereditary Cases

The three embryonal tumors under consideration all occur in both hereditary and nonhereditary forms (41,50,51). This was recognized earliest for retinoblastoma, because a greater number of survivors had children. It was also reported than when a survivor had had bilateral tumors, 50% of the offspring would be affected, regardless of previous family history. Since most bilateral cases do not have a previous family history, it has been concluded that most of them result from new germinal mutations rather than from transmitted mutations. Since the probability of retinoblastoma is 5×10^{-5} per child, since approximately 30% of cases are bilateral the inheritance of retinoblastoma is autosomal dominant, the rate for new germinal mutations is of the order of magnitude (5×10^{-5}) (0.3) (0.5), or approximately 7.5×10^{-6} per generation. The rate may be different because some bilateral cases are not the result of new germinal mutation but some unilateral cases are.

The location of the event has been determined as a consequence of a small fraction of cases that result from a constitutional deletion in chromosome 13. A few percent of these cases are heterozygous for a visible deletion that always includes a critical band, 13q14.1 (27,49,87). Some of the deletions have been so small that only this band is lost. Almost all of these deletions also involve hemizygosity for the enzyme esterase D, so the genetic loci are very close (82). In at least one case, a deletion did not include esterase D; it was therefore deduced that esterase D is slightly closer to the centromere than is the retinoblastoma locus (81).

Most patients with the hereditary form do not demonstrate such a deletion. In one case it was shown that esterase D was hemizygous even though no deletion was visible, a circumstance attributed to an invisible deletion (7). Since the human haploid genome contains approximately 3×10^9 base plairs, and since the maximum number of bands visualized in the haploid set is approximately 2,000, an occult deletion might still involve as many as 1,000 kilobases of DNA. Still remaining are the majority of hereditary cases with no evidence of deletion. The analysis of these cases has been possible because of the existence of a polymorphism for the electrophoretic variation of esterase. There are two major alleles, 1 and 2, giving three genotypes and phenotypes, 1/1, 1/2, and 2/2. Under certain circumstances, such as when a person with hereditary retinoblastoma is heterozygous for esterase D (i.e., type 1/2), and has affected and unaffected descendants, a linkage study can be performed. In two reported studies there have been no recombinants, indicating close proximity of the esterase D and retinoblastoma loci, just as in the deletion cases (18,80). It can therefore be concluded that hereditary retinoblastoma is attributable to a mutation of a gene located in band 13q14. In some cases, this mutation is a visible deletion; sometimes it is an invisible deletion; and sometimes another change is involved, which might include still smaller deletions or other submicroscopic lesions.

Such detailed information is not available for any other hereditary cancer; however, some cases of Wilms' tumor are localized cytogenetically. Some persons with the sporadic form of aniridia acquire Wilms' tumor (63). Following the proposal that these cases might be due to deletion (51), it was in fact demonstrated that deletion was the cause (26,57,75), the critical band being 11p13. What has not yet been demonstrated for Wilms' tumor is that the mutation in nondeletion hereditary cases is located at the same site.

Nonhereditary Cases

It was hypothesized that the hereditary and nonhereditary forms of a particular cancer begin with mutation at the same locus (41,42,50–52), the mutation being germinal in hereditary cases and somatic in nonhereditary cases. For both retinoblastoma (4,6,29) and Wilms' tumor (19,39,78,79), it has been found that approximately 20% of cases reveal an abnormality of the same chromosome as is affected in the hereditary cases. This percentage is much higher than for the hereditary cases, probably because loss of other genes on the same chromosome is more damaging to the whole organism than to just the retina; many constitutional deletions may be lethal.

Can we therefore predict that other hereditary cancers also involve mutation at genetic sites that undergo somatic mutation in the same form of that tumor? Unfortunately, other cancers have not been firmly localized, and this question cannot yet be answered.

Intermediate Cells

Since individuals with the hereditary form of a cancer develop only a few tumors, it is obvious that this event is not sufficient for oncogenesis. Nevertheless, there are several examples in such persons of abnormalities that have been viewed variously as developmental anomalies, hyperplastic lesions, or benign tumors. Examples are the neurofibromas seen in neurofibromatosis, the C-cell hyperplasia of the thyroid medulla seen in hereditary forms of medullary carcinoma of the thyroid (38), and the adenomatous polyps found in polyposis coli. For hereditary neurofibromas and colonic polyps (37), it has been ascertained that the lesions are not clonal in origin, suggesting that they have not resulted from a second genetic event. Evidently, the inherited mutation creates some abnormality that can be expressed under certain conditions. It may be that some cooperative mechanism is in operation. In normal persons, nonhereditary neurofibromas are clonal in origin, consistent with a somatic mutational first event, probably of the same gene affected in the hereditary cases.

The intermediate lesions predispose to malignancy, as is evident from the fact that the malignant tumors (e.g., carcinoma of the colon, neurofibrosarcoma) arise from them, not from cells outside the lesions. This may be due to their containing more mitotic target cells for transformation. In this sense, the lesions would resemble the papillomas that are produced in initiation–promotion experiments during chemical carcinogenesis in skin. The initiator produces an initiating mutation, the promoter, expansion of initiated cells into papillomas. Carcinomas are not the principal lesions produced in these experiments. One interpretation of this phenomenon was that formation of carcinomas required a second rare event that could be induced by further application of initiator after the promoter (65,73), which in fact has been shown to be the case (35).

Intermediate lesions have not been reported in many of the hereditary cancers, e.g., in retinoblastoma. For neuroblastoma it has been proposed that the tumors found in stage IV S in the skin, bone marrow, and liver are such lesions and that such patients are bearers of a germinal mutation (48). Consistent with this notion is the high rate of spontaneous regression of neuroblastoma IV S and the normal karyotypes of such tumors (3,10).

A Second, Genetic Event

In the hereditary cancers the mutant gene is transmitted dominantly, but only rare cells in the target tissue are transformed into tumors. Clearly, the inherited mutation is not sufficient for transformation. It was proposed that another genetic event was necessary, that the event involved loss or mutation of the homologous allele in the paired chromosome, and that the loss could occur as a new event (e.g., mutation, deletion, chromosome loss) or by somatic recombination to form a cell homozygous for the initial mutation (42,43,45). This latter suggestion was motivated by the observation of somatic recombination in Bloom's syndrome, a condition known to predispose to many forms of cancer, the idea being that these patients have an inherited ability to transform a somatically initiated heterozygous cell into a homozygous cell (24,43,71). For oncogenesis, the mutation was viewed as recessive.

If an event is a rare genetic change that occurs at approximately the same rate in different individuals, then it should be randomly distributed in a Poisson manner. In fact, enumeration of tumors in persons with bilateral retinoblastoma demonstrated a nearly Poisson distribution with a mean number of three (41).

Another consequence of a rare second tumor is that it should be derived from a single cell and be clonal with respect to an appropriate genetic marker. Such a marker is provided by the polymorphism at the X-linked locus for gluscoe-6-phosphate dehydrogenase (G6PD). In heterozygous females, only one allele is expressed in each cell, so a tumor derived from a single cell should express only one allele, whereas tissues generally express both. In hereditary cases of a tumor, the once-hit cell with the inherited mutation would express both G6PD alleles. Clonality has indeed been found for neurofibrosarcoma in patients with neurofibromatosis and for medullary carcinoma of the thyroid in patients with the hereditary form (5,28).

For retinoblastoma specifically, it was proposed that the tumor genotype could in theory be homozygously defective in any of six ways (45). If $13q^{rb}$ represents a submicroscopic mutation; $13q^-$, a visible deletion; and 13^-, loss of chromosome 13; then the six genotypes would be $13q^{rb}/13q^{rb}$, $13q^{rb}/13q^-$, $13q^{rb}/13^-$, $13q^-/13q^-$, and $13^-/13^-$. The latter two or three genotypes might produce nonviable phenotypes, however, so most tumors would be expected to be of the first three kinds.

It was further proposed that the requirement for a second genetic event applied to the nonhereditary form of a tumor as well as the hereditary form. By this formulation the mechanism of oncogenesis would be the same for the hereditary and nonhereditary forms of any tumor, the only difference being that the first event is germinal in the former case and somatic in the latter case, the second event being somatic in both. The same cancer can occur in either a normal $(+/+)$ individual or a heterozygous person $(+/-)$, the tumor being the same $(-/-)$ in both.

Theoretically, there should be a third group of individuals; those who are homozygous $(-/-)$ for the mutation (43,46). They would constitute one-quarter of the offspring of matings between two heterozygous individuals. Such individuals are not known. It might be that such a genotype would be lethal, with all cells in the target tissue becoming tumor cells during generation of the tissue in fetal life. If this were true, then the number of somatic events in oncogenesis would be 2, 1, or 0, according to whether the host genotype were $(+/+)$, $(+/-)$, or $(-/-)$.

Investigations of retinoblastoma and Wilms' tumor have indeed demonstrated that, at least in many cases, both normal alleles have been lost, and the genes are recessive in oncogenesis (13,20,23,33,54,69,70,74). In some cases, as predicted, the second event is a new one, such as deletion or chromosomal loss, whereas in others it is somatic recombination, with segregation of a cell homozygous for the initial mutation. Such findings have also been made for osteosarcoma (34), hepatoblastoma (53), and rhabdomyosarcoma (53). (Details are reviewed in the chapter by Benedict, *this volume*.)

Growth, Progression, and Additional Events

Other genetic events are known to occur in tumors, even in retinoblastoma, the classical example of the tumor attributable to a recessive cancer gene. Considerable controversy surrounds these events. Some, such as extensive hyperdiploidy, are probably nonspecific and are a reflection of the instability of the mitotic apparatus of the cancer cell; however, some are so specific that it is unlikely that they have no significance for tumor formation and/or growth. Thus, in retinoblastoma approximately 60% of tumors show a chromosome iso-6*p*, in addition to two normal chromosomes 6; a part of chromosome 6 occurs in four copies (6,56,83). The meaning of this is still not clear. One possibility is that it represents a step in tumor progression, not essential for formation of the tumor but causing the tumor to be more aggressive.

In 40% or so of the cases of neuroblastoma, there is amplification of the N-*myc* oncogene, and this change is related to metastatic properties (11). As many as 300 copies of the gene may be present. Some equally aggressive tumors do not have this change, so it cannot be viewed as essential for metastasis, but it is clearly important.

In the tumors of children the events are compressed into a brief time frame. Most cases of retinoblastoma, neuroblastoma, and Wilms' tumor are already apparent by the age of 6 years. For retinoblastoma and Wilm's tumor, loss of both normal alleles of a particular gene is essential for tumor formation. The number of rate-limiting events is two for the nonhereditary form and one for the hereditary form. Subsequent genetic changes, such as iso-6*p* formation or N-*myc* amplification, are evidently not rate-limiting. One explanation for this is that the first tumor cell results from the recessive state and that the tumor then grows in cell number to the point where further genetic changes are very probable. Thus, an event that occurs at a rate of 10^{-6} per cell division is rate-limiting for one cell, but highly probable and not rate-limiting for 10^6 cells.

A Mathematical Model

If two cellular events are rate-limiting for the development of a cancer, they should be the determinants of the age-specific incidence curve for a particular cancer. The "multihit" hypothesis thus becomes a "two-hit" hypothesis, whereas the hereditary forms of these tumors will behave as one-hit phenomena.

A mathematical two-event model has been constructed for retinoblastoma and shown to fit age-specific incidence data (36,47). In the hereditary form it can be shown that an assumption of rapid generation of retinoblasts followed by a stochastic event (a new mutation, deletion, or chromosomal loss; or recombination) and rapid growth to clinical detection is quite satisfactory. The rate of the second event would be of the order of magnitude of 10^{-7} per cell division among retinoblasts. Plotted in the form of yet undiagnosed cases versus age, the curve is that of a single event.

Any model for children's cancer must take into account the population of target cells as a function of time. These tumors (retinoblastoma, Wilms' tumor, neuroblastoma, rhabdomyosarcoma, medulloblastoma, etc.) all arise from cells that become postmitotic. The tissue progenitor cells become exhausted. The common feature of such populations is that the target cell number is zero at conception and again at some later age in childhood, with a maximum between these times. The events, or hits, that produce tumors must occur during the period when target cells are still present. Tumors that are common in the adult, on the other hand, arise in renewal tissues, so the population of target cells grows in early life as the tissue is generated and then remains constant, or nearly so. Tumors arising in such tissues could arise all through life. If more than one event is necessary, the age-specific incidence would be proportional to the integral of a function $X(s)$, where s is time, that describes the population of target cells (65,66).

We assume that the transition rate from target cell to intermediate cell is constant and is of the order of magnitude of a mutation rate, using that term to include all genetic change (point mutation, deletion, translocation, chromosomal loss, etc.). Of course, this rate could be increased above background levels by environmental agents, such as radiation, or by genetic abnormality, such as xeroderma pigmentosum. The age-specific incidence of the nonhereditary form of a tumor should be directly proportional to this rate.

Similar considerations apply to the second transition rate, except that it may include somatic recombination as well as mutation, if the two events affect the members of a gene pair. Again, environmental agents could increase the rate, although we should note the formal possibility that there may be agents that can increase the rate of genetic recombination in somatic cells without affecting mutation rate. Here, too, the incidence of a tumor should be proportional to the rate of this second event.

The kinetics of the intermediate cell evokes a different consideration. In the nonhereditary form, such cells will increase in number because their mitotic rate is greater than their rate of differentiation and death. As time passes, a singly mutant cell grows exponentially in proportion to the difference in rates, $\alpha - \beta$, and the interval of time, $t - s$, where t is a particular time and s is the time variable.

A general mathematical formulation of such a two-event (nonhereditary) cancer therefore assumes the following form (65):

$$I(t) = \text{age-specific incidence} = k\mu_1 \mu_2 \int_0^t X(s) \exp[(\alpha - \beta)(t - s)] \, ds$$

This model incorporates all of the elements noted above, namely, first-event rate (μ_1), second-event rate (μ_2), population of target cells [$X(s)$], and growth of intermediate cells $\{\exp[(\alpha - \beta)(t - s)]\}$. It is assumed that growth of tumor from one cell to clinical detection is not a rate-limiting step, an assumption that is required to render such a model useful. For many tumors this may be true, either in the sense that tumor growth requires less time than the formation of the first tumor cell, or that the time from one cell to detection is relatively constant. The observed incidences of several cancers have been shown to be fitted well by this equation.

NATURE OF THE EVENTS IN CARCINOGENESIS

Mutations in "Cancer Genes"

If the events discussed above are mutational, then our attention turns to the nature of the genes in which these mutations are occurring. In the case of retinoblastoma, it appears that the first step is *always* a mutation (including gross changes such as deletion) in a gene located in band 13q14.1. It is also likely that the second event leads to mutation or loss of the corresponding normal allele in the homologous chromosome. In such a situation, retinoblastoma may be viewed as a single gene disorder, similar to the recessively inherited inborn errors of metabolism, for example. A similar mechanism probably operates for Wilms' tumor. Evidence has been presented that suggests that osteosarcoma (34) and possibly alveolar rhabdomyosarcoma (77,84) similarly involve either the retinoblastoma gene itself or a very closely linked locus and that hepatoblastoma and embryonal rhabdomyosarcoma bear a similar relationship to Wilms' tumor (53). Two major questions arise immediately. Are these two events sufficient for carcinogenesis or are still other events necessary? This matter, discussed previously, is difficult to answer. The second question concerns the possibility that such a mechanism may operate in the common cancers of adults.

One candidate for such a mechanism in adults is renal carcinoma. A translocation between chromosomes 3 and 8 has been associated with this tumor in one family (16), the break points being at 3p14.2 and 8q24.1 (86). All persons in this family show either normal or translocation karyotypes, and the latter is highly penetrant for renal carcinoma. As with retinoblastoma, the number of primary tumors is very small, indicating that this translocation is not *sufficient* for carcinogenesis. In order to test whether this tumor may fit the retinoblastoma model, it will be necessary to determine whether a second event affects the homologous normal site and whether the tumors in nonhereditary cases show a deletion at the site. In one patient with the hereditary form, the constitutional karyotype was normal, but the tumor showed a break at 3p14 (72).

Since every cancer of adults exists in a hereditary form, it may be that this mechanism causes some fraction of all kinds of cancer, both hereditary and nonhereditary. It is likely that more than one genetic site is involved in some cancers. For example, two hereditary forms of colon carcinoma are known, one with polyposis and one without. The former affects primarily the left colon, the latter, the right colon, although the tumors are not separable pathologically. According to the recessive hypothesis, a second event mutates or removes the homologous normal allele. Furthermore, there should be two corresponding forms of nonhereditary colon carcinoma; one arises by somatic mutation in the polyposis coli locus of normal persons; the other in the family cancer syndrome locus. Similarly for breast cancer there are at least two hereditary forms; one affects women primarily beyond the age of 50 years and is sometimes associated with other cancers of adults; the other affects younger women and involves a gene that sometimes causes soft tissue sarcomas (59,61). There should be at least two nonhereditary forms of breast cancer therefore.

Should this prove to be correct, it would mean that there is a family of cancer genes, at least 50 in number, that plays a role in all cancer, both hereditary or nonhereditary and "spontaneous" or induced.

Oncogenes and Antioncogenes

The family of cancer genes discussed here is clearly different from the cellular oncogenes whose existence was revealed by the study of acutely transforming retroviruses. These oncogenes cause cancer in a positive sense, by producing an abnormal product, an abnormally large amount of product, or both. They can be transforming in the heterozygous state; i.e., they are dominant at the cellular level. In some human cancers, e.g., Burkitt's lymphoma, it appears that oncogene activation occurs via translocation that juxtaposes activating sequences and oncogene (76); however, this kind of translocation has never been found as a germinal mutation.

The recessive genes discussed above are quite different. Here, both alleles must be abnormal or absent. Cancer is evidently caused by the absence of a normal product. Viewed another way, the normal allele is antioncogenic. For this reason, such genes have been called antioncogenes to distinguish them from oncogenes (46).

So it seems that two different classes of genes may be involved in the origin of human cancer. Are these truly different mechanisms, or is there some final common pathway? Two classes of explanation of the latter type can be mentioned. One would be that both kinds of change are necessary in every tumor, the other that the same final effect is produced by an interaction of the two kinds of genes.

It cannot be ruled out yet that both kinds of change are found in all tumors. Even in a condition like Burkitt's lymphoma it is possible that an antecedent change has occurred (40). Thus, the translocation in chronic myeloid leukemia, resulting in the Philadelphia chromosome, appears to have been preceded by a clonal change of a different kind in at least some cases (25). On the other hand, tumors like retinoblastoma may require some other genetic event, such as oncogene aberration, before their malignancy is fully expressed. Therein could lie the significance of such changes as the iso-6p chromosome in that tumor.

The other kind of explanation entails an interaction between the two classes of genes. One scenario calls for the regulation of the oncogene by diploid regulatory genes, the recessive genes already discussed (17). Cancer could result from an unresponsive heterozygous alteration in an oncogene or homozygous loss of the normal regulatory alleles. In retinoblastoma, e.g., there is elevated expression of the N-*myc* oncogene, possibly the result of lost regulatory sites (58).

Another kind of interaction calls for the recessive gene coding for an important component of a differentiated cell, such as a cell surface protein. Loss of both alleles would result in loss of this protein. This model would be similar to that of one form of hypercholesterolemia, in which there is loss of lipoprotein receptors (32). Conversely, impairment of the activity of this protein might result from interaction with an abnormal oncogene product.

Alternative to such additive or interactive explanations is the possibility that the two mechanisms of oncogenesis are quite independent—that some tumors involve one, and some the other.

Cancer Genes: Development and Oncogenesis

The effects of mutations in these two classes of genes can be dramatic. It is possible that most human cancers result from such mutations. Questions now arise with regard to the normal function of such genes.

One long-standing view of cancer is that it is a form of abnormal development. It is therefore hardly surprising that these cancer genes are suspected of being important in such development. Support for this idea with respect to oncogenes comes from their differential tissue expression during embryogenesis (67). So far, teratogenic effects caused by mutations of oncogenes have not been observed, although peculiar tumors of the choroid plexus have been observed following transgenic passage of a mutant oncogene in mice (9).

There is more evidence that the recessive genes are important in development. The occurrence of germinal mutations of these genes has permitted observations about their tissue specificities and effects on tissue development. All of the mutations show some specificity in heterozygotes, and many show considerable specificity. For example, the gene of Sipple's disease predisposes to pheochromocytoma and medullary carcinoma of the thyroid. This disease, in fact, is one of at least 10 hereditary conditions that effect tumors in derivatives of the neural crest. Thus, there are additional mutations for pheochromocytoma alone and medullary carcinoma alone, for neuroblastoma, for melanoma (the dysplastic nevus syndrome), for chemodectoma, for meningioma, for acoustic neuroma (central form of neurofibromatosis), for neurofibrosarcoma (peripheral form of neurofibromatosis), and for mucosal neuroma (multiple endocrine neoplasia type 3) (46). These conditions suggest that there are numerous genes that are important for normal histogenesis.

In some instances, even the heterozygous state for these recessive cancer genes produces observable effects in tissues. The polyps in polyposis coli and the neurofibromas in neurofibromatosis, both sometimes classified as hamartomatous lesions, are examples. In the heritable form of medullary carcinoma of the thyroid, C-cell hyperplasia is regularly found in the thyroid (38), and persistent renal blastema is sometimes associated with Wilms' tumor (8).

A conclusive demonstration of the importance of these genes in development would depend on the existence of individuals homozygous for the mutation. In the absence of the normal allele one might observe developmental aberrations. Persons of such genotype have not been observed, but there are two circumstances in animals that suggest that development defects might occur. In the rat, there is a mutation for which the heterozygous state strongly predisposes to renal carcinoma of the type seen in humans (21,22). Homozygous mutant animals have not been recovered from matings between heterozygotes, implying that the recessive genotype is lethal before birth. The author (A.G. Knudson, Jr., *unpublished data*) has found that 25 percent of the embryos of such matings die at day 9 to 10. There is also a mutation in *Drosophila melanogaster* that may be relevant. This mutation gives rise to a neuroblastic tumor in the optic disc of the recessive larva, resulting in death of the larva (30). In both the rat and the fruit fly, lack of the normal allele interferes with normal embryonic development and leads to tumor. In the fly, it has been observed that neuroblasts in the third instar of normal larvae divide unequally, one daughter cell becoming another neuroblast, the other entering a path that leads to terminal differentiation. In the mutant fly, mitosis gives rise to two neuroblasts that go on to form tumors.

CONCLUSIONS

The incidences of hereditary and nonhereditary cancers in children can be explained by a two-event model, according to which the two events are considered to lead to the homozygous loss of the normal alleles of one of a set of "cancer genes"; i.e., the mutations are recessive in oncogenesis. We have cited several cytogenetic and molecular studies that strongly support such a model.

The common cancers of adults also occur in both hereditary and nonhereditary form and their incidence can be fitted to a generalized two-event model so that they, too, may result from the homozygous loss of normal alleles of this class of genes. Use of molecular probes of polymorphic DNA fragments may confirm the existence of such genes for adult cancers.

These genes, which are distinct from oncogenes, have been called antioncogenes. Major questions in carcinogenesis concern the importance of these two classes of genes in cancer generally. Is one class important for some cancers and the second for other cancers, or are both important for most cancers? Is there a physiological relationship between the two classes of genes?

The normal alleles of both oncogenes and antioncogenes may be critical during development. The lethality of recessive genes in the rat and fruit fly suggests that at least normal antioncogenes are significant developmental genes and that oncogenesis and development are closely related, as has long been thought.

ACKNOWLEDGMENT

Preparation of this manuscript was supported by Grant CA-06927 from the United States Public Health Service and by an appropriation from the Commonwealth of Pennsylvania.

REFERENCES

1. Armitage, P., and Doll, R. (1954): The age distribution of cancer and a multistage theory of carcinogenesis. *Br. J. Cancer*, 8:1–12.
2. Armitage, P., and Doll, R. (1957): A two-stage theory of carcinogenesis in relation to the age distribution of human cancer. *Br. J. Cancer*, 11:161–169.
3. Balaban, G., and Gilbert, F. (1983): Neuroblastoma IV-S: chromosome analysis. *N. Engl. J. Med.*, 309:989.
4. Balaban, G., Gilbert, F., Nichols, W., Meadows, A.T., and Shields, J. (1982): Abnormalities of chromosome No. 13 in retinoblasts from individuals with normal constitutional karyotypes. *Cancer Genet. Cytogenet.*, 6:213–221.
5. Baylin, S.B., Hsu, S.H., Gann, D.S., Smallridge, R.C., and Wells, S.A. (1978): Inherited medullary thyroid carcinoma: a final monoclonal mutation in one of multiple clones of susceptible cells. *Science*, 199:429–431.
6. Benedict, W.F., Banerjee, A., Mark, C., and Murphree, A.L., (1983): Non-random chromosomal changes in untreated retinoblastomas. *Cancer Genet. Cytogenet.*, 10:311–333.
7. Benedict, W.F., Murphree, A.L., Banerjee, A., Spina, C.A., Sparkes, M.D., and Sparkes, R.S. (1983): Patient with 13 chromosome deletion: evidence that the retinoblastoma gene is a recessive cancer gene. *Science*, 219:973–975.
8. Bove, K.E., and McAdams, A.J. (1976): The nephroblastomatosis complex and its relationship to Wilms' tumor: a clinico-pathologic treatise. In: *Perspectives in Pediatric Pathology*, Vol. 3, pp. 185–223. Year Book Medical Publishers, Chicago.
9. Brinster, R.L., Chen, H.Y., Messing, A., van Dyke, T., Levine, A.J., and Palmiter, R.D. (1984): Transgenic mice harboring SV40 T-antigen genes develop characteristic brain tumors. *Cell*, 37:367–379.
10. Brodeur, G., Green, A.A., Hayes, F.A., Williams, K.J., Williams, D.L., and Tsiatis, A.A. (1981): Cytogenetic features of human neuroblastomas and cell lines. *Cancer Res.*, 41:4678–4686.
11. Brodeur, G.M., Seeger, R.C., Schwab, M., Varmus, H.E., and Bishop, J.M. (1984): Amplification of N-*myc* in untreated human neuroblastomas correlates with advanced disease stage. *Science*, 224:1121–1124.
12. Burch, P.R. (1965): Natural and radiation carcinogenesis in man. I. Theory of initiation phase. *Proc. R. Soc. Lond.*, B162:223–239.
13. Cavenee, W.K., Dryja, T.P., Phillips, R.A., Benedict, W.F., Godbout, R., Gallie, B.L., Murphree, A.L., Strong, L.C., and White, R.L. (1983): Expression of recessive alleles by chromosomal mechanisms in retinoblastoma. *Nature*, 305:779–784.
14. Chaganti, R.S., Schonberg, S., and German, J. (1974): A manyfold increase in sister chromatid exchanges in Bloom's syndrome lymphocytes. *Proc. Natl. Acad. Sci. U.S.A.*, 71:4508–4512.
15. Cleaver, J.E., and Bootsma, D. (1975): Xeroderma pigmentosum: biochemical and genetic characteristics. *Annu. Rev. Genet.*, 9:19–38.
16. Cohen, A.J., Li, F.P., Berg, S., Marchetto, D.J., Tsai, S., Jacobs, S.C., and Brown, R.S. (1979): Hereditary renal-cell carcinoma associated with a chromosomal translocation. *N. Engl. J. Med.*, 301:592–595.
17. Comings, D.E. (1973): A general theory of carcinogenesis. *Proc. Natl. Acad. Sci. U.S.A.*, 70:3324–3328.
18. Connolly, M.J., Payne, R.H., Johnson, G., Gallie, B.L., Allerdice, P.W., Marshall, W.H., and Lawton, R.D. (1983): Familial, EsD-linked, retinoblastoma with reduced penetrance and variable expressivity. *Hum. Genet.*, 65:122–124.
19. Douglass, E.C., Williams, J.A., Green, A.A., and Look, A.T. (1985): Abnormalities of chromosomes 1 and 11 in Wilms' tumor. *Cancer Genet. Cytogenet.*, 14:331–338.
20. Dryja, T.P., Cavenee, W., White, R., Rapaport, J.M., Peterson, R., Albert, D.M., and Bruns, G.A.P. (1984): Expression of recessive alleles by chromosomal mechanisms in retinoblastoma. *N. Engl. J. Med.*, 310:550–553.
21. Eker, R., and Mossige, J. (1961): A dominant gene for renal adenomas in the rat. *Nature*, 189:858–859.
22. Eker, R., Mossige, J., Johannessen, J.V., and Aars, H. (1981): Hereditary renal adenomas and adenocarcinomas in rats. *Diag. Histopathol.*, 4:99–110.

23. Fearon, E.R., Vogelstein, B., and Feinberg, A.P. (1984): Somatic deletion and duplication of genes on chromosome 11 in Wilms' tumours. *Nature*, 309:176–178.
24. Festa, R.S., Meadows, A.T., and Boshes, R.A. (1979): Leukemia in a black child with Bloom's syndrome. *Cancer*, 44:1507–1510.
25. Fialkow, P.J., Martin, P.J., Najfeld, V., Penfold, G.K., Jacobson, R.J., and Hansen, J.A. (1981): Evidence for a multistep pathogenesis of chronic myelogenous leukemia. *Blood*, 58:158–163.
26. Francke, U., Holmes, L.B., Atkins, L., and Riccardi, V.M. (1979): Aniridia-Wilms' tumor association: evidence for specific deletion of 11p13. *Cytogenet. Cell Genet.*, 24:185–192.
27. Francke, U., and Kung, F. (1976): Sporadic bilateral retinoblastoma and 13q^- chromosomal deletion. *Med. Pediatr. Oncol.*, 2:379–385.
28. Friedman, J.M., Fialkow, P.J., Greene, C.L., and Weinberg, M.N. (1983): Probable clonal origin of neurofibrosarcoma in a patient with hereditary neurofibromatosis. *J. Natl. Cancer Inst.*, 69:1289–1292.
29. Gardner, H.A., Gallie, B.L., Knight, L.A., and Phillips, R.A. (1982): Multiple karyotypic changes in retinoblastoma tumor cells: presence of normal chromosome No. 13 in most tumors. *Cancer Cenet. Cytogenet.*, 6:201–211.
30. Gateff, E. (1978): Malignant neoplasms of genetic origin in *Drosophila melanogaster*. *Science*, 200:1448–1459.
31. German, J. (1974): Bloom's syndrome. II. The prototype of human genetic disorders predisposing to chromosome instability and cancer. In: *Chromosomes and Cancer.* edited by J. German, pp. 601–617. John Wiley and Sons, New York.
32. Goldstein, J.L., and Brown, M.S. (1979): The LDL receptor locus and the genetics of familial hypercholesterolemia. *Annu. Rev. Genet.*, 13:259–289.
33. Godbout, R., Dryja, T.P., Squire, J., Gallie, B.L., and Phillips, R.A. (1983): Somatic inactivation of genes on chromosome 13 is a common event in retinoblastoma. *Nature*, 304:451–453.
34. Hansen, M.F., Koufos, A., Gallie, B.L., Phillips, R.A., Fodstad, O., Brogger, A., Gedde-Dahl, T., and Cavenee, W.K. (1985): Osteosarcoma and retinoblastoma: A shared chromosomal mechanism revealing recessive predisposition. *Proc. Natl. Acad. Sci. U.S.A.*, 82:1–5.
35. Hennings, H., Shores, R., Wenk, M.L., Spangler, E.F., Tarone, R., and Yuspa, S.H. (1983): Malignant conversion of mouse skin tumours is increased by tumour initiators and unaffected by tumour promoters. *Nature*, 304:67–69.
36. Hethcote, H.W., and Knudson, A.G. (1978): Model for the incidence of embryonal cancers: Application to retinoblastoma. *Proc. Natl. Acad. Sci. U.S.A.*, 75:2453–2457.
37. Hsu, S.H., Luk, G.D., Krush, A.J., Hamilton, S.R., and Hoover, H.H. (1983): Multiclonal origin of polyps in Gardner syndrome. *Science*, 221:951–953.
38. Jackson, C.E., Block, M.A., Greenawald, K.A., and Tashigian, A.H. (1979): The two-mutational-event theory in medullary thyroid cancer. *Am. J. Hum. Genet.*, 31:704–710.
39. Kaneko, Y., Egues, M.C., and Rowley, J.D. (1981): Interstitial deletion of short arm of chromosome 11 limited to Wilms' tumor cells in a patient without aniridia. *Cancer Res.*, 41:4577–4578.
40. Klein, G., and Klein, E. (1985): Evolution of tumours and the impact of molecular oncology. *Nature*, 315:190–195.
41. Knudson, A.G. (1971): Mutation and cancer: statistical study of retinoblastoma. *Proc. Natl. Acad. Sci. U.S.A.*, 68:820–823.
42. Knudson, A.G. (1973): Mutation and human cancer. *Adv. Cancer Res.*, 17:317–352.
43. Knudson, A.G. (1976): Germinal and somatic mutations in cancer. In: *Excerpta Medica International Congress Series No. 411*, pp. 367–371. (Proceedings of the Fifth International Congress of Human Genetics Mexico City, Oct. 10–15, 1976.) Excerpta Medica, Amsterdam.
44. Knudson, A.G. (1977): Genetics and etiology of human cancer. *Adv. Hum. Genet*, 8:1–66.
45. Knudson, A.G. (1978): Retinoblastoma: a prototypic hereditary neoplasm. *Semin. Oncol.*, 5:57–60.
46. Knudson, A.G. (1985): Hereditary cancer, oncogenes, and antioncogenes. *Cancer Res.*, 45:1437–1443.
47. Knudson, A.G., Hethcote, H.W., and Brown, B.W. (1975): Mutation and childhood cancer: A probabilistic model for the incidence of retinoblastoma. *Proc. Natl. Acad. Sci. U.S.A.*, 72:5116–5120.
48. Knudson, A.G., and Meadows, A.T. (1980): Regression of neuroblastoma IV-S: A genetic hypothesis. *N. Engl. J. Med.*, 302:1254–1256.

49. Knudson, A.G., Meadows, A.G., Nichols, W.W., and Hill, R. (1976): Chromosomal deletion and retinoblastoma. *N. Engl. J. Med.*, 295:1120–1123.
50. Knudson, A.G., and Strong, L.C. (1972): Mutation and cancer: neuroblastoma and pheochromocytoma. *Am. J. Hum. Genet.*, 24:514–532.
51. Knudson, A.G., and Strong, L.C. (1972): Mutation and cancer: a model for Wilms' tumor of the kidney. *J. Natl. Cancer Inst.*, 48:313–324.
52. Knudson, A.G., Strong, L.C., and Anderson, D.E. (1973): Heredity and cancer in man. In: *Progress in Medical Genetics*, edited by A.G. Steinberg and A.G. Bearn, Vol IX, pp. 113–158. Grune and Stratton, New York.
53. Koufos, A., Hansen, M.F., Copeland, N.G., Jenkins, N.A., Lampkin, B.C., and Cavenee, W.K. (1985): Loss of heterozygosity in three embryonal tumours suggest a common pathogenetic mechanism. *Nature*, 316:330–334.
54. Koufos, A., Hansen, M.F., Lampkin, D.B., Workman, M.L., Copeland, N.G., Jenkins, N.A., and Cavenee, W.K. (1984): Loss of alleles at loci on human chromosome 11 during genesis of Wilms' tumour. *Nature*, 309:170–172.
55. Kraemer, K.H., (1980): Oculo-cutaneous and internal neoplasms in xeroderma pigmentosum: implications for theories of carcinogenesis. In: *Carcinogenesis: Fundamental Mechanisms and Environmental Effects*, edited by B. Pullman, P.O.P. Ts'o, and H. Gelboin, pp. 503–507. D. Reidel Publishing Company, Holland.
56. Kusnetsova, L.E., Prigogina, E.L., Pogosianz, H.E., and Belkina, B.M. (1982): Similar chromosomal abnormalities in several retinoblastomas. *Hum. Genet.*, 61:201–204.
57. Ladda, R., Atkins, L., Littlefield, J., Neurath, P., and Marimuthu, K.M. (1974): Computer-assisted analysis of chromosomal abnormalities: detection of a deletion in aniridia/Wilms' tumor syndrome. *Science*, 185:784–787.
58. Lee, W.-H., Murphree, A.L., and Benedict, W.F. (1984): Expression and amplification of the N-*myc* gene in primary retinoblastoma. *Nature*, 309:458–460.
59. Li, F.P., and Fraumeni, J.F. (1975): Familial breast cancer, softtissue sarcomas, and other neoplasms. *Ann. Intern. Med.*, 83:833–834.
60. Lynch, H.T., Frichot III, B.C., and Lynch, J.F. (1977): Cancer control in xeroderma pigmentosum. *Arch. Dermatol*, 113:193–195.
61. Lynch, H.T., Guirgis, H.A., Brodkey, F.D., Lynch, J., Maloney, K., Rankin, L., and Mulcahy, G.M. (1976): Genetic heterogeneity and familial carcinoma of the breast. *Surg. Gynecol. Obstet.*, 142:693–699.
62. Maher, V.M., Ouellette, L.M., Curren, R.D., and McCormick, J.J. (1976): Frequency of ultraviolet light-induced mutation is higher in xeroderma pigmentosum variant cells than in normal human cells. *Nature*, 261:593–595.
63. Miller, R.W., Fraumeni, J.F., and Manning, M.D. (1964): Association of Wilms' tumor with aniridia, hemihypertrophy and other congenital malformations. *N. Engl. J. Med.*, 270:922–927.
64. Mintz, B., and Sanyal, S. (1970): Clonal origin of the mouse visual retina mapped from genetically mosaic eyes. *Genetics*, 64:43–44.
65. Moolgavkar, S.H., and Knudson, A.G. (1981): Mutation and cancer: a model for human carcinogenesis. *J. Natl. Cancer Inst.*, 66:1037–1052.
66. Moolgavkar, S.H., and Venzon, D.J. (1979): Two-event models for carcinogenesis: Incidence curves for childhood and adult tumors. *Math. Biosci.*, 47:55–77.
67. Muller, R., Tremblay, J.M., Adamson, E.D., and Verma, I.M. (1983): Tissue and cell type-specific expression of two human c-*onc* genes. *Nature*, 304:454–456.
68. Mulvihill, J.J. (1971): Genetic reportory of human neoplasia. In: *Genetics of Human Cancer*, edited by J.J. Mulvihill, R.W. Miller, and J.F. Fraumeni, pp. 137–143. Raven Press, New York.
69. Murphree, A.L., and Benedict, W.F. (1984): Retinoblastoma: clues to human oncogenesis. *Science*, 223:1028–1033.
70. Orkin, S.H., Goldman, D.S., and Sallan, S.E. (1984): Development of homozygosity for chromosome 11p markers in Wilms' tumor. *Nature*, 309:172–174.
71. Passarge, E., and Bartram, C.R. (1976): Somatic recombination as possible prelude to malignant transformation. In: *Cancer and Genetics*, edited by D. Bergsma, Vol. 12, pp. 177–180. March of Dimes, Alan R. Liss, Inc. New York.
72. Pathak, S., Strong, L.C., Ferrell, R.E., and Trindade, A. (1982): Familial renal cell carcinoma with a 3;11 chromosome translocation limited to tumor cells. *Science*, 217:939–941.
73. Potter, V.R. (1981): A new protocol and its rationale for the study of initiation and promotion of

carcinogenesis in rat liver. *Carcinogenesis*, 2:1375–1379.
74. Reeve, A.E., Housiaux, P.J. Gardner, R.J.M., Chewings, W.E., Grindley, R.M., and Millow, L.J. (1984): Loss of Harvey *ras* allele in sporadic Wilms' tumour. *Nature*, 309:174–176.
75. Riccardi, V.M., Hittner, H.M., Francke, U., Yunis, J.J., Ledbetter, D., and Borges, W. (1980): The Aniridia-Wilms' tumor association: the clinical role of chromosome band 11p13. *Cancer Genet. Cytogenet.*, 2:131–137.
76. Rowley, J.D. (1984): Biological implications of consistent chromosome rearrangements in leukemia and lymphoma. *Cancer Res.*, 44:3159–3168.
77. Seidal, T., Mark, J., Hagmar, B., and Angervall, L. (1982): Alveolar rhabdomyosarcoma: a cytogenetic and correlated cytological and histological study. *Acta Path. Microbiol. Immunol. Scand. Sect A*, 90:345–354.
78. Slater, R.M. and de Kraker, J. (1982) Chromosome number 11 and Wilms' tumor. *Cancer Genet. Cytogenet.*, 5:237–245.
79. Slater, R.M., de Kraker, J., Voute, P.A., and Delemarre, J.F.M. (1985): A cytogenetic study of Wilms' tumor. *Cancer Genet. Cytogenet.*, 14:95–109.
80. Sparkes, R.S., Murphree, A.L., Lingua, R.W., Sparkes, M.C., Field, L.L., Funderburk, S.J., and Benedict, W.F. (1983): Gene for hereditary retinoblastoma assigned to human chromosome 13 by linkage to esterase D. *Science*, 219:971–973.
81. Sparkes, R.S., Sparkes, M.C., Kalina, R.E., Pagon, R.A., Salk, D.J., and Disteche, C.M. (1984): Separation of retinoblastoma and esterase D loci in a patient with sporadic retinoblastoma and del(13)(q14.1q22.3). *Hum. Genet.*, 68:258–259.
82. Sparkes, R.S., Sparkes, M.C., Wilson, M.G., Towner, J.W., Benedict, W., Murphree, A.L., and Yunis, J.J. (1980): Regional assignment of genes for human esterase D and retinoblastoma to chromosome band 13q14. *Science*, 208:1042–1044.
83. Squire, J., Phillips, R.A., Boyce, S., Godbout, R., Rogers, B., and Gallie, B.L. (1984): Isochromosome 6p, a unique chromosomal abnormality in retinoblastoma: verification by standard staining techniques, new densitometric methods, and somatic cell hybridization. *Hum. Genet.* 66:46–53.
84. Turc-Carel, C., and Philip, T. (1985): Consistent chromosomal translocation t(2;13)(q37;q14) in alveolar rhabdomyosarcoma. In: *Proceedings XVIIth Meeting of International Society of Pediatric Oncology, Venice, Italy, Sep. 30–Oct. 4, 1985*.
85. Vijayalaxmi, Evans, H.J., Ray, J.H., and German, J. (1983): Bloom's syndrome: evidence for an increased mutation frequency in vivo. *Science*, 221:851–853.
86. Wang, N., and Perkins, K.L. (1984): Involvement of band 3p14 in t(3:8) hereditary renal carcinoma. *Cancer Genet. Cytogenet.*, 11:479–481.
87. Yunis, J.J., and Ramsay, N. (1978): Retinoblastoma and subband deletion of chromosome 13. *Am. J. Dis. Child.*, 132:161–163.

Recessive Human Cancer Susceptibility Genes (Retinoblastoma and Wilms' Loci)

William F. Benedict

Division of Ophthalmology and the Clayton Ocular Oncology Center, Childrens Hospital of Los Angeles, Los Angeles, California 90027

At present, there appear to be two major classes of potential human cancer genes. The first are oncogenes, which apparently play a role in tumor formation of nonhuman cells by their aberrant activation and expression. Mechanisms for this activation include somatic mutations, increased expression of a single-copy gene, and gene amplification. Thus far, the evidence suggests that the expression of oncogenes is dominant.

In contrast, there are definite human cancer susceptibility genes that may produce tumors following their inactivation or loss of expression. Such inactivation could occur by deletion (either microscopic or submicroscopic), mutation, or chromosomal translocation into the gene itself. This class of genes is apparently recessive, and although they have been termed "antioncogenes" we prefer the term "suppressor" or "regulatory" genes (38). The reason for taking this position is that no definitive evidence has been obtained indicating that these alleles have any relationship to the expression of oncogenes, although this may be an attractive hypothesis. Moreover, the term antioncogene also has been used to indicate that the normal gene is antioncogenic (28), which also adds to the confusion in the terminology.

Recessive cancer susceptibility genes appear to have a particularly important role in the development of tumors arising in early childhood or adolescent years; however, they also may be responsible for tumors in adults. Considerable information has been published on two such human cancer susceptibility genes. The first gene is located at the "retinoblastoma" or Rb locus (Rb-1). In addition to its role in retinoblastoma formation, this locus is also involved in the development of osteosarcoma (1,16,23). The second susceptibility gene is located at "Wilms' tumor" or WAGR locus, which has been implicated in the formation of hepatoblastoma and rhabdomyosarcoma (31) as well as Wilms' tumor.

RETINOBLASTOMA LOCUS

Historical Perspective

In 1971, Al Knudson (27) presented his hypothesis for the development of retinoblastoma. This hypothesis was based on statistical analysis of clinical data, which included the fact that patients with the dominantly inherited bilateral form of retinoblastoma were diagnosed at an earlier age than those individuals with the nonhereditary unilateral form of the disease. Knudson postulated that in bilateral retinoblastoma, a first "hit" or mutation occurred at the germinal level. He proposed that two mutations were required for tumor development and that the earlier diagnosis of the sporadic bilateral form of retinoblastoma resulted from the fact that a first "mutation" had already occurred in each target cell. Thus, only one additional hit was required for tumor formation. In the unilateral, nonhereditary form of retinoblastoma, both mutations were thought to occur at the somatic level in a single "retinoblast" resulting in a longer latency period (27).

David Comings (10) 2 years later proposed a more general theory of tumorigenesis in which he incorporated elements of the oncogene theory of George Todaro and Robert Huebner (50) and the "two-hit" theory of Al Knudson (27). Comings suggested that in every cell there are structural "transforming" genes that are active normally during embryogenesis. He proposed that these genes were later suppressed by diploid pairs of "suppressor" or "regulatory" alleles during differentiation. Comings also postulated that a mutation in both of these regulatory alleles could result in the release of suppression. This in turn would allow the transforming gene to be expressed continuously, resulting in cellular transformation. Results from cytogenetic studies (2–5,9) and molecular biological analysis (7,8,15) suggest that both Knudson's and Comings' hypotheses indeed may have considerable merit.

Loss of One Rb-1 Allele as the "First Event" in the Development of Retinoblastoma

A small proportion of patients with the hereditary form of retinoblastoma (initially estimated to be approximately 5%) were known to have a constitutional deletion of chromosome 13. Yunis and Ramsay (54) later confirmed by deletion mapping and high-resolution chromosomal banding analysis that the common deletion in all of these patients included chromosomal region $13q14$. This finding showed, at least in the 13 deletion cases, that the first event was a total loss of one Rb allele. Motegi (35,36) then proposed that there may be an even larger number of patients with retinoblastoma who have a mosiacism including a $13q14$ deletion as the first event in retinoblastoma development. She reported that three of 15 patients examined using high-resolution banding techniques showed a constitutional mosaicism including a $13q14$ deletion (36). If confirmed by other laboratories, this observation would imply that the first event initially postulated by Knudson (27) and expanded by Comings (10) is frequently a visible structural deletion of a suppressor or regulatory gene.

Localization of the Rb-1 Locus to 13q14.11

Until 1983, some individuals speculated that the locus for the nondeletion form of hereditary retinoblastoma was not located on chromosome 13 (33). Fortunately, during this year, the issue was resolved (46). The gene locus for the polymorphic enzyme, esterase D (EsD), also previously had been assigned to chromosomal region 13q14 by deletion mapping (48). Since EsD activity is gene dose dependent, 50% EsD activity is usually present in the constitutional cells of patients with the deletion form of retinoblastoma (48). In addition, the fact that EsD is polymorphic allowed us to determine in patients with the nondeletion familial form of retinoblastoma that the same Rb-1 locus is responsible for the susceptibility to develop tumors. Because of the tight linkage between the EsD and Rb-1 loci, it was possible to show that either the EsD1 or EsD2 allele segregated with the development of retinoblastoma in specific families (46). Such an example is illustrated in Fig. 1. The EsD2 allele segregates with the development of retinoblastoma in this particular family. From these studies, we concluded that there was a common locus in 13q14 responsible for both the hereditary deletion and nondeletion forms of retinoblastoma (46). Similar results showing the inheritance of a specific EsD allele and the susceptibility to retinoblastoma were also reported by others (11).

Subsequently, using high-resolution chromosomal banding, Motegi et al. (37) were able to assign both the Rb and EsD loci to 13q14.1. Even more recently, both loci have been mapped to 13q14.11 (51). The fact that the EsD and Rb-1 loci are indeed distinct was shown in a patient with retinoblastoma who had a large constitutional deletion including 13q14.1, but in whom 100% EsD activity was present (47). These results indicate that the breakpoint in this individual was between the EsD and the Rb-1 loci and implies that the EsD locus is located closer to the centromere than the Rb-1 locus.

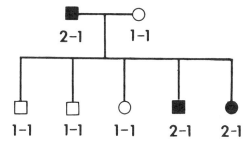

Fig. 1. Pedigree of a family with the hereditary nondeletion form of retinoblastoma. Individuals with retinoblastoma are shown in black, and whether they had a 1-1 or 2-1 electrophoretic pattern for EsD is shown below. In this family, the inheritance of the EsD2 allele was linked to the development of retinoblastoma. Thus, the chromosome 13 with the EsD allele carries the Rb allele as well. (From ref. 46.)

Recessive Nature of the Rb Gene

The recessive nature of the Rb gene has been discussed in detail by us (38), as well as others (28,41). A chronological discussion of the evidence leading to this conclusion shall be presented here.

In 1981, Louise Strong and her colleagues (49) published an extensive pedigree in which retinoblastoma occurred over four generations and was transmitted by eight unaffected individuals. Cytogenetic analysis of the pedigree showed that the inheritance of retinoblastoma was associated with a constitutional deletion including 13q14. Individuals who were unaffected carriers had a balanced insertional translocation, including 13q14 present in one chromosome 3. Thus, although the carriers had a deleted chromosome 13, the presence of one extra copy of this chromosomal 13 material (specifically, 13q14) translocated to chromosome 3 was sufficient to protect these individuals from developing retinoblastoma. This study inferred that the Rb locus was recessive at the cellular level, although the susceptibility for tumor formation is dominantly inherited.

Subsequently, a patient who exemplifies the recessive nature of the Rb-1 locus and may be key to understanding the function of the Rb gene was described by us in 1983 (5). She was found to have 50% EsD activity in her constitutional cells, including red blood cells, lymphoblasts, and fibroblasts. This patient therefore was considered to have the 13 deletion form of retinoblastoma; however, when her constitutional cells were examined at the 450 band level, no chromosomal deletion was found (Fig. 2). Jorge Yunis (53) also reported that there was no deletion present at the 2,000 band level in this patient's lymphoblasts. Based on this information, we suggest that this individual may have a submicroscopic deletion within 13q14.11 in which both the EsD and Rb-1 locus have been lost.

Next, we examined the tumor from the one enucleated eye from this patient to determine its chromosomal pattern. Two distinct tumor cell lines were observed by karyotypic analysis, both of which had a total loss of one 13 chromosome (5). A metaphase from one stemline is shown in Fig. 3.

It then became important to determine whether the normal or submicroscopically deleted 13 chromosome had been lost in the tumor lines. Therefore, the EsD activity was measured in the tumor, and no activity was found on several determinations (5). We concluded that it was the *normal* 13 chromosome that had been lost in the tumor and postulated that the total loss of both Rb-1 alleles was necessary for the development of retinoblastoma (5,38). These findings strongly supported the recessive nature of the Rb-1 locus.

At first, there appears to be a paradox in terminology, since the susceptibility to retinoblastoma is autosomal dominantly inherited. However, it is the fact that the second event apparently will occur in one of the potential target cells with greater than 90% certainty that makes the inheritance to develop a tumor dominant.

Further support that a structural or functional loss of both Rb-1 alleles plays a fundamental role in the development of retinoblastoma comes from the chromo-

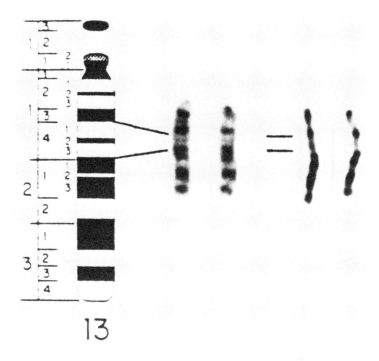

Fig. 2. Lack of visable deletion of chromosome 13 in the patient with 50% EsD activity described in text at the 450 band level. Diagram of the banding pattern of chromosome 13 is shown at the far left. The inserted lines bracket chromosomal region 13q14. All subbands (q14.1, q14.2, and q14.3) were present in both fibroblasts (*center*) and lymphoblasts (*right*) examined from this patient. (From ref. 5.)

somal analysis of tumors. In our study of direct tumor preparations from 20 individual tumor stemlines, we found that 25% had a total loss of one chromosome 13, with an additional tumor having a chromosomal deletion including 13q14 (4). previously, Gloria Balaban and her colleagues (2,3) also had found patients with a deletion including 13q14 in several retinoblastomas. We think that the normal Rb allele was lost in the bilateral cases, although there is no method available at present to prove this contention.

At the same time that the cytogenetic studies were being undertaken by us, Webb Cavenee et al. (6) and Ted Dryja et al. (14) were isolating various restriction-fragment-length polymorphisms (RFLPs) located on chromosome 13. Using these probes, Cavenee and his colleagues (7) investigated the status of each chromosome 13 in various retinoblastomas. They first confirmed that there was a total loss of one 13 chromosome in the patient's tumor whose karyotype is shown in Fig. 3 (7). Additional tumors in which there were at least two normal-appearing chromosomes 13 by cytogenetic analysis were found to be homozygous for all or a portion of one chromosome 13, including the Rb-1 locus (7). It was concluded

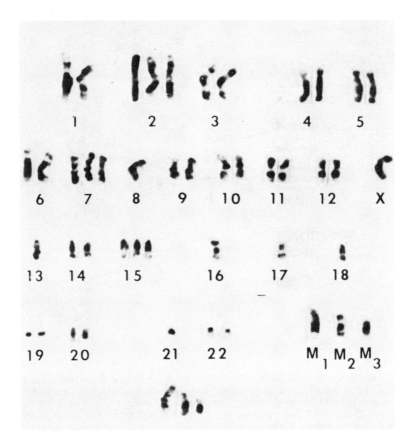

Fig. 3. Karyotype showing a loss of one chromosome 13 in a tumor cell from the patient with a submicroscopic constitutional deletion of chromosome 13. (From ref. 5.)

that either a mitotic nondisjunction with reduplication of the remaining chromosome or a mitotic recombination occurred, which presumably resulted in the loss of all of a portion of the homologous wild-type chromosome, particularly at the Rb-1 locus (7). These findings may explain the loss of heterozygosity for one EsD allele found in four of six tumors from patients who were constitutionally heterozygous for the alleles (22). Based on these results, Godbout and her associates (22) also proposed that the induction of tumor formation required the somatic inactivation of genes close to the EsD locus, including the normal allele at the Rb-1 locus. Subsequently, Dryja et al. (15) also reported results similar to those of Cavenee and his colleagues (6).

It was not possible to prove with certainty in any of the above cases that homozygosity for the initially "mutant" Rb-1 locus had occurred, since the patients studied did not have the familial form of retinoblastoma (7,15); however, two patients with familial retinoblastoma have been examined, and indeed their tumors showed homozygosity for the mutant recessive allele at the Rb-1 locus (8). Thus, it is likely that the homozygosity seen in the other tumors also involve the mutant allele.

In collaboration with Webb Cavenee and Mark Hansen, we also have examined six of our tumors from patients with both unilateral and bilateral disease. All tumors contained two normal-appearing chromosomes 13, and the patients were constitutionally heterozygous for various RFLPs on chromosome 13. Five of six tumors showed homozygosity for all or a portion of one chromosome 13, which apparently resulted from a nondisjunction and reduplication or a mitotic recombination involving chromosome 13 (*unpublished results*).

Depicted in Fig. 4 are several mechanisms that would produce hemizygosity or homozygosity at the Rb-1 locus. The first possiblity is a total loss of the chromosome containing the wild-type Rb^+ allele by nondisjunctional loss. This would lead to hemizygosity for the mutant allele and result in a nonrandom chromosome 13 loss. Such a situation has been observed in several of the primary tumors that we examined (4). The second possibility shown (Fig. 4b) again involves a nondisjunctional loss of the normal wild-type allele with reduplication of the mutated chromosome. A mitotic recombination (Fig. 4c) would also lead to homozygosity for the Rb^- allele. Since additional RFLPs will be required to distinguish between b and c, it can only be stated at present that one or both of these mechanisms appears to be the most common second event (probably greater than 50%) leading to homozygosity at the Rb-1 locus (7,14).

The fourth possibility for the second event (Fig. 4d) is a deletion in the chromosome containing the wild-type allele that includes chromosomal region $13q14$. Such a deletion has been seen by ourselves (4) as well as by others (2,3,9) in various tumors. What we believe to be a very rare occurrence is shown as mechanism e, i.e., gene inactivation. By this terminology, we refer to the translocation of the normal chromosome 13, which would include the wild-type Rb^+ allele, to an inactive X chromosome. In this situation, the structural Rb^+ allele is present but would be no longer functional. Such an event has been reported as the first constitutional event in patients having retinoblastoma (12,17,24). We have found one case in whch such a mechanism apparently has occurred as a second event (W.F. Benedict, *unpublished results*). Finally, there could be a point or frameshift mutation in the Rb^+ allele (f). Until the Rb gene is isolated, it will not be possible to prove that such an event actually occurs. At present, we consider that a mutation could be involved as the first event more frequently in tumor development than as a second event, since mechanisms a–d shown in Fig. 4 appear to represent the vast majority of second events.

Development of Osteosarcoma Involves the Rb-1 Locus

It is well known that patients with hereditary retinoblastoma have a high risk for developing second nonocular tumors. The most common of these malignancies is osteosarcoma. David Abramson and his colleagues (1) reported in a large series

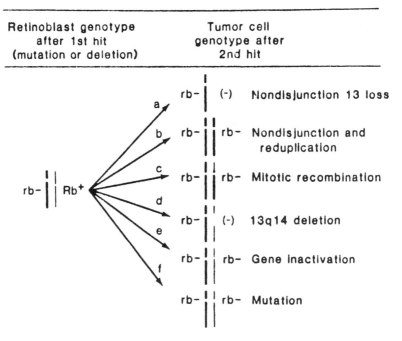

Fig. 4. Mechanisms producing hemizygosity or homozgosity at the Rb-1 locus. A complete discussion of these mechanisms is outlined in the text. (from ref. 38). We have assumed that a mutation or deletion (either microscopic or submicroscopic) has occurred in one retinoblast as the "first hit." The aberrant allele is depicted as Rb$^-$ to reflect the recessive nature of this locus as well as our present thinking that there is actually a loss or functional inactivation at this allele (5,38). The normal allele at the Rb-1 locus is depicted as Rb$^+$, indicating its dominant effect over the "mutated" allele at the Rb-1 locus. To the right are shown various second events that could occur leading to tumor formation.

that more than 50% of the individuals who survived hereditary retinoblastoma had developed a second primary malignancy, particularly osteosarcoma, within 30 years of the diagnosis of retinoblastoma. We have proposed, therefore, that the Rb gene must be considered a potent human cancer susceptibility gene that can affect more than one target tissue (38).

Only recently, however, has evidence been obtained at the molecular level that suggests that a similar mechanism occurs in the development of both osteosarcoma and retinoblastoma. When RFLPs were used on chromosome 13, two of three second primary osteosarcomas were found to have lost their constitutional heterozygosity for chromosome 13, including the Rb-1 locus (23). Perhaps even more remarkable was the fact that three of four osteosarcomas from patients without any history of retinoblastoma likewise specifically lost their constitutional heterozygosity for various loci along chromosome 13 (23). An additional report (16) showing complete homozygosity of chromosome 13 in three osteosarcoma lines from patients without a history of retinoblastoma is consistent with this generalized mechanism of tumor development.

These results indicate that the development of homozygosity at the Rb-1 locus is likely to be an important factor in the development of osteosarcoma. In addition, they suggest that there may be a limited number of loci where recessive "mutations" can predispose to malignancy, since these loci could have a broad spectrum of tissue specificity for producing cancer.

WILMS' TUMOR LOCUS

There are numerous similarities between the development of Wilms' tumor and retinoblastoma. In 1964, Robert Miller and his colleagues (34) reported an association between the development of Wilms' tumor and aniridia. Subsequently, Al Knudson and Louise Strong (29) proposed that the basis for Wilms' tumor was similar to that for retinoblastoma. Similar to the situation for retinoblastoma, individuals with bilateral Wilms' tumor were considered to be hereditary cases. Individuals in which aniridia was associated with Wilms' tumor were hypothesized to have a constitutional deletion, including genes responsible for both aniridia and Wilms' tumor (29).

It took several years before cytogenetic studies revealed a common deletion in these patients (20,43,44), which confirmed this hypothesis (29). Only recently, has the exact location of the (WAGR) Wilms' tumor, aniridia, genitourinary abnormalities, and mental retardation (WAGR) locus has been mapped by high-resolution banding techniques (39). Subsequently, homozygosity of the WAGR locus was shown to be highly frequent in Wilms' tumor (18,31,40,42), similar to the changes reported for the Rb-1 locus in the induction of retinoblastoma (7,15). Finally, the same mechanism of tumor formation, again involving the WAGR locus, has been implicated as the basis for rhabdomyosarcoma and hepatoblastoma (30). These findings and the similarity to the development of retinoblastoma are now to be discussed in detail.

Loss of One WAGR Allele (11*p*1305 to *p*1306) as the "First Event" in the Development of Wilms' Tumor

A small number of patients with the hereditary form of Wilms' tumor (between 1% and 3%) have a constitutional deletion in chromosome 11. This deletion was initially mapped to 11*p*13 (20,43) and more recently has been assigned to chromosome region 11*p*1305 to *p*1306 (39). The gene coding for catalase (CAT) has also been mapped to this same region (19,21,25,39,52) and is more distally located to the centromere than the WAGR locus (39). Unfortunately, CAT in contrast to EsD is not polymorphic, and thus it has not been possible to determine with certainty that the nondeletion hereditary form of Wilms' tumor also involves the WAGR locus; however, this likely is the case.

Recessive Nature of the Wilms' Tumor Gene

There are numerous parallels between the development of retinoblastoma and Wilms' tumor that indicate that the mechanism of tumorigenesis is similar. The deletion of 11p13 in patients with the aniridia, a hereditary form of Wilms' tumor (20,39,43), parallels the deletion of 13q14 in patients with the hereditary, deletion form of retinoblastoma (54). In addition, a family has been described in which there was an interstitial deletion including 11p13 in affected family members, whereas unaffected individuals who were carriers had an additional interstitial translocation including chromosomal region 11p13 (55). This family is comparable to the pedigree reported by Strong et al. (49) in whom an insertional translocation including 13q14 was sufficient to prevent the development of retinoblastoma. Therefore, this study implies the recessive nature of the WAGR locus (55).

Patients without a constitutional deletion also on occasion have been found to have a visible deletion of 11p13 in their tumor (13,26,45). Such studies indicate that a deletion in 11p13 can be a "second event" in the formation of Wilms' tumor that is comparable to individuals with retinoblastoma who have a deletion in 13q14 (2-4,9).

The most compelling evidence for the similarity of the mechanism of tumor formation between Wilms' tumor and retinoblastoma comes from molecular biological studies. Four papers were published simultaneously that reported that hemizygosity or homozygosity for chromosome 11 frequently occurred in Wilms' tumor (18,31,40,42). These studies strongly suggest that there are recessive "mutational" events that become hemizygous or homozygous on abnormal chromosomal segregation or mitotic recombination during mitosis. Such mechanisms are similar to that illustrated in Fig. 4b and c for the role of the Rb-1 locus in the development of retinoblastoma.

Rhabdomyosarcoma and Hepatoblastoma Involves the WAGR Locus

Children with the autosomal dominantly inherited Beckwith-Wiedemann syndrome (BWS) have a markedly increased potential to develop three embryonal tumors: Wilms' tumor, hepatoblastoma, and rhabdomyosarcoma. Since the WAGR locus on human chromosome 11 is involved in the development of Wilms' tumor, the logical extension of this fact would be to examine the role of chromosome 11 in the development of hepatoblastoma and rhabdomyosarcoma. Such a study has been reported by Alex Koufos and his colleagues (30) using RFLPa located on chromosome 11. They documented that several loci on chromosomal arm 11p became homozygous or hemizygous, compared to their constitutional pattern in two of three embryonal hepatoblastomas. These results therefore were similar to those reported for Wilms' tumor (18,31,40,42); however, since neither case was familial, it was not possible to determine with certainty that the mutant allele on chromosome 11 became homozygous.

In the same report (31), two of three embryonal rhabdomyosarcomas were also found to be homozygous for loci onchromosome 11. These results again implicated a mitotic recombination or nondisjunction, and reduplication occurred in one chromosome 11. They also indicate that the number of recessive alleles that predispose to cancer may be fewer than originally thought. It is perhaps more likely that a specific locus has a broad but specific tissue type for which it can play a role in tumor formation.

Koufos et al. (30) also suggested an approach to localize other recessive alleles that cause cancer. In patients with autosomal dominant syndromes associated with certain malignancies, it may be possible to determine that a particular chromosome undergoes a loss of constitutional heterozygosity by using various RFLPs located on specific chromosomes. As proposed by Koufos et al. (31), these could include such conditions as neurofibromatosis, multiple endocrine neoplasia, and the basal cell nevus syndrome which are associated with astrocytoma, medullary thyroid carcinoma, and medulloblastoma, respectively. Such studies should in turn provide information on additional recessive cancer susceptibility loci similar to the Rb-1 and the WAGR loci and expand the role of these genes in future years.

Additional Considerations with Regard to the Rb-1 Locus and WAGR Loci

Critical information is still missing concerning the proposed first and second events at the Rb-1 and WAGR loci responsible for tumor development. These data are needed not only to help resolve questions remaining about the function of these loci, but also to define approaches that appear most likely to enable the cloning of these genes.

We have suggested that there may be a total structural or functional loss of both Rb alleles as the basis for tumor formation (38). This hypothesis was based primarily on the findings that the first event in certain patients included a deletion of the Rb-1 locus. Moreover, the patient previously discussed, with an apparent submicroscopic deletion including both the Rb-1 and EsD locus (Fig. 2), had lost the normal chromosome 13 in the tumor, suggesting that there was no structural gene at the Rb-1 locus in these particular tumor cells (5). However, several possibilities still exist for the development of retinoblastoma or osteosarcoma, as diagrammed in Fig. 5 that also could occur in Wilms' tumor development.

Depicted are two situations that could have occurred in either the "retinoblast" or the "osteoblast" target cell as the first event in tumor formation. The first is a recessive mutation, as shown in Fig. 5A. Subsequently, hemizygosity for the initial mutation could be produced by loss of the normal chromosome 13 (not shown). Homozygosity for this same mutation could be achieved either by reduplication of the mutated chromosome following the loss by nondisjunction of the normal chromosome (Fig. 5A-1) or by a mitotic recombination in which a portion of the normal chromosome not including the Rb-1 locus is retained (Fig. 5A-2). If

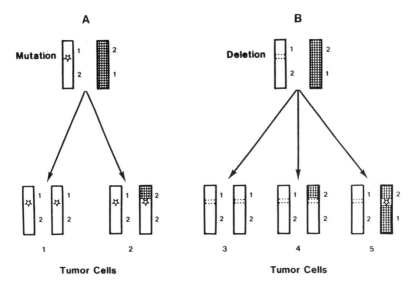

Fig. 5. Possibilities for tumor development of both retinoblastoma and osteosarcoms. Either a recessive mutation in one chromosome 13 shown in **A** as a *star* or a microscopic or submicroscopic deletion depicted in **B** as a *dotted line* could occur as the first event in tumor development. The normal chromosome 13 or chromosomal segment is represented by *small squares*. Two hypothetical RFLP probes, one above and one below the Rb-1 locus, are also shown to enable a given chromosome segment to be followed more clearly. Specific discussion of these possibilities is given in the text.

indeed the initial event is a recessive mutation that encodes for an aberrant gene product, then this gene locus could function in tumor formation by allowing, for example, the expression of an anomolous protein or the production of an abnormal receptor.

If, in contrast, the initial first event is a deletion or total inactivation of the Rb-1 locus (Fig. 5B), then several other alternatives exist. The first is the development of hemizygosity for the deletion by the loss of the normal chromosome 13 (not depicted). Homozygosity at this deletion could also occur either by subsequent reduplication of the deleted chromosome 13 (Fig. 5B-3) or by a mitotic recombination (Fig. 5B-4). Unfortunately, there have been no definitive data obtained on either possibility, since tumors from patients with constitutional deletions have not yet been examined in detail by high-resolution cytogenetic techniques, measurement of EsD activity, or study of RFLPs located on chromosome 13. If homozygosity for the deleted region could be shown, this would be consistent with our proposal that a structural or functional loss at the Rb-1 locus may be responsible for tumor development (5,38).

A final important mechanism for tumor formation could exist however, in which, following a deletion of one Rb allele, a recessive mutation in the second Rb allele occurs (Fig. 5B-5). Again, this mutation could result in the expression of an aberrant recessive gene product similar to that outlined in Fig. 5A-1 or 5A-2.

It is apparent to us, because of the absence of definitive data to distinguish among the various possibilities, that emphasis should be placed on obtaining tumors from patients with known constitutional deletions of the Rb-1 locus. We could then attempt to document the loss of the remaining normal Rb allele in such tumors.

If the loss of both Rb alleles can be documented, a group of approaches would be suggested that more likely might enable the successful cloning of the Rb gene and provide an understanding of its function. This would include DNA subtraction techniques to obtain the smallest segment of DNA missing in the tumors from patients with the deletion form of retinoblastoma for further cloning of the Rb gene (30). In addition, two-dimensional gel electrophoresis could be utilized to look for reexpression of a normal protein in the tumor compared to adult retinal cells. If, in contrast, the expression of a recessive mutation at the Rb-1 locus is responsible for tumor formation, then different approaches might be more successful. A discussion of these alternatives, however, is beyond the scope of this article.

Information similarly needs to be obtained on the function of the WAGR locus comparable to that outlined in Fig. 5 for the Rb-1 locus. Since only rare cases of Wilms' tumor have a deletion within 11p13 as the first event in tumor development, it is imperative that tumors from such patients be examined to determine if mechanisms occur that are depicted in Fig. 5B-3 and 5B-4 for the Rb-1 locus. Should such events actually be documented, it would indicate that a total loss of function in both alleles at the WAGR locus also may be the basis for tumor development.

There is presently considerable effort being expended in several laboratories, including our own, to clone both the Rb-1 and WAGR loci and to understand their functions. It is our hope that new molecular biological techniques combined with more standard approaches, such as the use of cell hybrids, will enable the rapid elucidation of how these recessive human cancer susceptibility loci contribute to tumor development.

ACKNOWLEDGMENTS

I wish to thank Drs. Linn Murphree, Webb Cavenee, Robert Sparkes, and Ashutosh Bannerjee for their key roles in various studies. The careful review of this manuscript by Drs. Garrett Brodeur and Alfred Knudson was greatly appreciated. Finally, I also wish to thank Corey Mark for his excellent technical assistance and Suzanne Bence for her expertise in preparing the manuscript. This research was supported in part by NIH Grant EYO-2715 and was done in conjunction with the Clayton Foundation for Research.

REFERENCES

1. Abramson, D.H., Ellsworth, R.M., Kitchin, F.D., and Tung, G. (1984): Second monocular tumors in retinoblastoma survivors. *Ophthalmology*, 91:1351–1355.
2. Balaban, G., Gilbert, F., Nichols, W., Meadows, A.T., and Shields, J. (1982): Abnormalities of

chromosome #13 in retinoblastomas from individuals with normal constitutional karyotypes. *Cancer Genet. Cytogenet.*, 6:213–221.
3. Balaban-Malenbaum, G., Gilbert, F., Nichols, W.W., Hill, R., Shields, J., and Meadows, A.T. (1981): A deleted chromosome No. 13 in human retinoblastoma cells: Relevance to tumorigenesis. *Cancer Genet. Cytogenet.*, 3:243–250.
4. Benedict, W.F., Banerjee, A., Mark, C., and Murphree, A.L. (1983): Non-random chromosomal changes in direct preparations of primary retinoblastoma. *Cancer Genet. Cytogenet.*, 10: 311–333.
5. Benedict, W.F., Murphree, A.L. Banerjee, A., Spina, C.A., and Sparkes, M.D., Sparkes, R.S. (1983): Evidence from a patient with the deletion form of retinoblastoma that the retinoblastoma gene is a recessive cancer gene. *Science*, 219:973–975.
6. Cavenee, W., Leach, R., Mohandas, T., Pearson, P., and White, R. (1984): Isolation and regional localization of DNA segments revealing polymorphic loci from human chromosome 13. *Am. J. Hum. Genet.*, 36:10–24.
7. Cavenee, W.K., Dryja, T.P., Phillips, R.A., Benedict, W.F., Godbout, R., Gallie, B.L., Murphree, A.L., Strong, L.C., and White, R.L. (1983): Expression of recessive alleles by chromosomal mechanisms in retinoblastoma. *Nature*, 305:779–784.
8. Cavenee, W.K., Hansen, M.F., Nordenskjold, M., Kock, E., Maumenee, I., Squire, J.A., Phillips, R.A., and Gallie, B.L. (1985): Genetic origin of mutations predisposing to retinoblastoma. *Science*, 228:501–503.
9. Chaum, E., Ellsworth, R.M., Abramson, D.H., Haik, B.G., Kitchin, F.D., and Chaganti, R.S.K. (1984): Cytogenetic analysis of retinoblastoma: evidence for multifocal origin and *in vivo* gene amplification. *Cytogenet. Cell Genet.*, 38:82–91.
10. Comings, D.E. (1973): A general theory of carcinogenesis. *Proc. Natl. Acad. Sci. U.S.A.*, 70:3324–3328.
11. Connolly, M.J., Payne, R.H., Johnson, G., Gallie, B.L., Allerdice, P.W., Marshall, W.H., and Lawton, R.D. (1983): Familial, EsD-linked, retinoblastoma with reduced penetrance and variable expressivity. *Hum. Genet.*, 65:122–124.
12. Cross, H.E., Hansen, R.C., Morrow III, G., and Davis, J.R. (1977): Retinoblastoma in a patient with a 13qXp translocation. *Am. J. Ophthalmol*, 84:548–554.
13. Douglass, E.C., Wilimas, J.A., Green, A.A., and Look, A.T. (1985): Abnormalities of chromosomes 1 and 11 in Wilms' tumor. *Cancer Genet. Cytogenet.* 14:331–338.
14. Dryja, T.P., Bruns, G.A.P. Orkin, S.H., Albert, D.M., and Gerald, P.S. (1983): Isolation of DNA fragments from chromosome 13. *Retina*, 3:121–125.
15. Dryja, T.P., Cavenee, W., White, R., Rapaport, J.M., Petersen, R., Albert, D. M., and Bruns, G.A.P. (1984): Homozygosity of chromosome 13 in retinoblastoma. *N. Engl. J. Med.*, 310:550–553.
16. Dryja, T.P., Rapaport, J.M., Epstein, J., Goorin, A.M., Weichselbaum, R., Koufos, A., and Cavenee, W.K. (1986): Chromosome 13 homozygosity in osteogenic sarcoma without retinoblastoma. *Am. J. Hum. Genet.*, 38:59–66.
17. Ejima, Y., (Sasaki, M.S., Kaneko, A., Tanooka, H., Hara, Y., Hida, T., and Kinoshita, Y. (1982): Possible inactivation of part of chromosome 13 due to 13qXp translocation associated with retinoblastoma. *Clin. Genet.*, 21:357–361.
18. Fearon, E.R., Vogelstein, B., and Feinberg, A.P. (1984): Somatic deletion and duplication of genes on chromosome 11 in Wilms' tumours. *Nature*, 309:176–178.
19. Ferrell, R.E., and Riccardi, V.M. (1983): Catalase levels in patients with aniridia and/or Wilms' tumor: Utility and limitations. *Cytogenet. Cell Genet.*, 31:120–123.
20. Francke, U., Holmes, L.B., Atkins, L., and Riccardi, V.M. (1979): Aniridia-Wilms' tumor association: Evidence for specific deletion of 11p13. *Cytogenet. Cell Genet.*, 24:185–192.
21. Gilgenkrantz, S., Vigneron, C., Gregoire, M.J., Pernot, C., and Raspiller, A. (1982): Association of del(11)(p15.1p12), aniridia, catalase deficiency, and cardiomyopathy. *Am. J. Med. Genet.*, 13:39–49.
22. Godbout, R., Dryja, T.P., Squire, J., Gallie, B.L., and Phillips, R.A. (1983): Somatic inactivation of genes on chromosome 13 is a common event in retinoblastoma. *Nature*, 304:451–453.
23. Hansen, M.F., Koufos, A., Gallie, B.L., Phillips, R.A., Fødstad, Ø., Brøgger, A., Gedde-Dahl, T., and Cavenee, W.K. (1985): Osteosarcoma and retinoblastoma: A shared chromosomal mechanism revealing recessive predisposition. *Proc. Natl. Acad. Sci. U.S.A.*, 82:6216–6220.
24. Hida T., Kinoshita, Y., Matsumoto, R., Suzuki, N., and Tanaka, H. (1980): Bilateral retinoblastoma with a 13qXp translocation. *J. Pediatr. Ophthal. Strabismus*, 17:144–146.

25. Junien, C. Turleau, C., Grouchy, J. de, Said, R., Rethore, M.O., Tenconi, R., and Dufier, J.L. (1980): Regional assignment of catalase (CAT) gene to band 11p13: Association with the aniridia-Wilms tumor-gonadoblastoma (WAGR) complex. *Ann. Genet.*, 29:165–168.
26. Kaneko, Y., Egues, M.C., Rowley, J.D. (1981): Interstitial deletion of short arm of chromosome 11 limited to Wilms' tumor cells in a patient without aniridia. *Cancer Res.*, 41:4577–4578.
27. Knudson, A.G. (1971): Mutation and cancer: Statistical study of retinoblastoma. *Proc. Natl. Acad. Sci. USA*, 68:820–823.
28. Knudson, A.G., (1985): Hereditary cancer, oncogenes, and antioncogenes. *Cancer Res.*, 45:1437–1443.
29. Knudson, A.G., and Strong, L.C. (1972): Mutation and cancer: A model for Wilms' tumor of the kidney. *J. Natl. Cancer Inst.*, 48:313–324.
30. Koufos, A., Hansen, M.F., Copeland, N.G., Jenkins, N.A., Lampkin, B.C., and Cavenee, W.K. (1985): Loss of heterozygosity in three embryonal tumours suggests a common pathogenetic mechanism. *Nature*, 316:330–334.
31. Koufos, A., Hansen, M.F., Lampkin, B.C., Workman, M.L., Copeland, N.G., Jenkins, N.A., and Cavenee, W.K. (1984): Loss of alleles at loci on human chromosome 11 during genesis of Wilms' tumor. *Nature*, 309:170–172.
32. Kunkel, L.M., Monaco, A.P., Middlesworth, W., Ochs, H.D., and Latt, S.A. (1985): Specific cloning of DNA fragments absent from the DNA of a male patient with an X chromosome deletion. *Proc. Natl. Acad. Sci. U.S.A.*, 82:4778–4782.
33. Matsunaga, E. (1980): Retinoblastoma: Host resistance and $13q^-$ chromosomal deletion. *Hum. Genet.*, 56:53–58.
34. Miller, R.W., Fraumeni, J.F., and Mannin, M.D. (1964): Association of Wilms tumor with aniridia, hemihypertrophy, and other congenital malformations. *N. Engl. J. Med.*, 270:922–927.
35. Motegi, T. (1981): Lymphocyte chromosome survey in 42 patients with retinoblastoma: Effort to detect 13q14 deletion mosaicism. *Hum. Genet.*, 58:168–173.
36. Motegi, T. (1982): High rate of detection of 13q14 deletion mosaicism among retinoblastoma patients (using more extensive methods). *Hum. Genet.*, 61:95–97.
37. Motegi, T., Komatsu, M., Nakazato, Y., Ohuchi, M., Minoda, K. (1982): Retinoblastoma in a boy with a de novo mutation of a 13/18 translocation: The assumption that the retinoblastoma locus is at 13q141, particularly at the distal portion of it. *Hum. Genet.*, 60:193–195.
38. Murphree, A.L., and Benedict, W.F. (1984): Retinoblastoma: Clues for human oncogenesis. *Science*, 223:1028–1033.
39. Narahara, K., Kikkawa, K., Kimira, S., Kimoto, H., Ogata, M., Kasai, R., Hamawaki, M., and Matsuoka, K. (1984): Regional mapping of catalase and Wilms tumor-aniridia, genitourinary abnormalities, and mental retardation triad loci to the chromosome segment 11p1305 → p1306. *Hum. Genet.*, 66:181–185.
40. Orkin, S.H., Goldman, D.S., and Sallan, S.E. (1984): Development of homozygosity for chromosome 11p markers in Wilms' tumor. *Nature*, 309:172–174.
41. Phillips, R.A., and Gallie, B.L. (1984): Retinoblastoma: Importance of recessive mutations in tumorigenesis. *J. Cell. Physiol. [Suppl.]*, 3:79–85.
42. Reeve, A.E., Housiaux, P.J., Gardner, R.J., Chewings, W.E., Grindley, R.M., and Millow, L.J. (1984): Loss of a Harvey *ras* allele in sporadic Wilms' tumour. *Nature*, 309:174–176.
43. Riccardi, V.M., Hittner, H.M., Francke, U., Yunis, J.J., Ledbetter, D., and Borges, W. (1980): The aniridia-Wilms' tumor association: The critical role of chromosome band 11p13. *Cancer Genet. Cytogenet.*, 2:131–138.
44. Riccardi, V.M., Sujansky, E., Smith, A.C., and Francke, U. (1978): Chromosomal imbalance in the aniridia-Wilms' tumor association: 11p interstitial deletion. *Pediatrics*, 61:604–610.
45. Slater, R.M., and de Kraker, J. (1982): Chromosome number 11 and Wilms' tumor. *Cancer Genet. Cytogenet.*, 5:237–245.
46. Sparkes, R.S., Murphree, A.L., Lingua, R.W., Sparkes, M.C., Field, L.L., Funderburk, S.J., and Benedict, W.F. (1983): Gene for hereditary retinoblastoma assigned to human chromosome 13 by linkage to esterase D. *Science*, 219:917–979.
47. Sparkes, R.S., Sparkes, M.C., Kalina, R.E., Pagon, R.A., Salk, D.J., and Disteche, C.M. (1984): Separation of retinoblastoma and esterase D loci in a patient with sporadic retinoblastoma and del(13)(q14.1q22.3). *Hum. Genet.*, 68:258–259.
48. Sparkes, R.S., Sparkes, M.C., Wilson, M.G., Towner, J.W., Benedict, W.F., Murphree, A.L., and Yunis, J.J. (1980): Regional assignment of genes for human esterase D and retinoblastoma to chromosome band 13q14. *Science*, 208:1042–1043.

49. Strong, L.C., Riccardi, V.M, Ferrell, R.E., and Sparkes, R.S. (1981): Familial retinoblastoma and chromosome 13 deletion transmitted via an insertional translocation. *Science*, 213:1501–1503.
50. Todaro, G.J., and Huebner, R.J. (1972): *Proc. Natl. Acad. Sci. U.S.A.*, 69:1009–1015.
51. Ward, P., Packman, S., Loughman, W., Sparkes, M., Sparkes, R., McMahon, A., Gregory, T., and Ablin, A. (1984): Location of the retinoblastoma susceptibility gene(s) and the human esterase D locus. *J. Med. Genet.*, 21:92–95.
52. Wieacker, P., Muller, C.R., Mayer, A., Grzeschik, K.H., and Ropers, H.H. (1980): Assignment of the gene coding for humn catalase to short arm of chromosome 11. *Ann. Genet. (Paris)*, 23:73–77.
53. Yunis, J.J. (1983): The chromosomal basis of human neoplasia. *Science*, 221:227–236.
54. Yunis, J.J., and Ramsay, N. (1978): Retinoblastoma and subband deletion of chromosome 13. *Am. J. Dis. Child.*, 132:161–163.
55. Yunis, J.J., and Ramsay, N.K.C. (1980): Familial occurrence of aniridia-Wilms tumor syndrome with deletion 11p13-14.1. *J. Pediat.*, 96:1027–1030.

Molecular Basis of Human B- and T-Cell Neoplasia

*Carlo M. Croce, *Jan Erikson, *Yoshihide Tsujimoto,
and **Peter C. Nowell

*The Wistar Institute, Philadelphia, Philadelphia 19104; and **Department of Pathology and Laboratory Medicine, University of Pennsylvania School of Medicine, Philadelphia, Pennsylvania 19104

Most human leukemias and lymphomas carry nonrandom chromosomal alterations, predominantly translocations and inversions (1). Following the discovery of the Philadelphia chromosome in chronic myelogenous leukemia (2), it was recognized that specific chromosomal rearrangements are associated with certain human malignancies. These changes were considered by many to be an epiphenomenon and not the cause of human neoplasia. Only after the discovery that in Burkitt's lymphoma the human homologue, c-*myc*, of the avian myelocytomatosis virus oncogene, v-*myc* (3), and that in chronic myelogenous leukemia the human homologue, c-*abl*, of the Abelson leukemia virus oncogene v-*abl* (4), are directly involved in the specific chromosome translocations observed in these malignancies did it become clear that specific chromosomal rearrangements play an important role in the pathogenesis of human hematopoietic neoplasms.

BURKITT'S LYMPHOMA

In Burkitt's lymphoma, a malignant condition of B cells that affects predominantly children, Manolov and Manolova (5) detected a chromosome 14 ($14q^+$) larger than normal. Such a chromosome was found by Zech et al. (6) to be the result of a translocation of the distal end of the long arm of chromosome 8 to chromosome 14. This t(8;14) chromosome translocation was found in approximately 75% of human Burkitt's lymphoma. During the past 6 years, two variant chromosomal translocations have been detected in Burkitt's lymphoma (7,8). In one case, the distal end of the long arm ($q24$-$q$ter) of chromosome 8 translocates to the long arm of chromosome 22, whereas in the other case, the same region of chromosome 8 translocates to the short arm of chromosome 2 (7,8). Approximately 16% of Burkitt's lymphoma carries the t(8;22)(q24;q11) translocation, whereas the remaining 9% carries the t(2;8)(p11;q24) chromosome translocation (7,8). Since we have previously mapped the heavy-chain locus to chromosome 14 (9) and the λ light-chain locus to chromosome 22 (10), we postulated that the human immunoglobulin genes may play an important role in the pathogenesis of Burkitt's lymphoma (10). In view of the chromosomal location of the human immunoglobulin genes, Klein (11) postulated that the human immunoglobulin genes could play an important role in the pathogenesis of Burkitt's lymphoma by providing a cellular oncogene with immunoglobulin promoters. These suggestions were strengthened by the mapping of the κ chain locus to the short arm of chromosome 2 by McBride et al. (12) and Malcom et al. (13). The proof that human immunoglobulin loci are directly involved in the rearrangements observed in Burkitt's lymphoma was provided by the analysis of somatic cell hybrids between mouse myeloma cells and Burkitt's lymphoma cells carrying the t(8;14)(q24;q32) chromosome translocation (14). Since it was observed that hybrids containing the $14q^+$ chromosome retain the genes for the constant regions of the heavy chains, whereas hybrids containing the $8q^-$ chromosome retain the genes for the variable regions of the heavy chains, it was concluded that the chromosome breaks on chromosome 14 in Burkitt's lymphoma involve directly the heavy-chain locus (14) (Fig. 1). It is interesting to note that it was also found that the normal chromosome 14 carries the productively rearranged heavy-chain locus (14).

The next breakthrough came from the chromosome mapping of the human homologues of retroviral oncogenes (3). We have found that the human c-*myc* oncogene translocated to the heavy-chain locus in Burkitt's lymphoma carrying the t(8;14) chromosome translocation (3). Dalla Favera et al. (3) and others (15,16) also found that in some Burkitt's lymphoma the translocated c-*myc* gene is rearranged, whereas in others, it is in its germ line configuration. On the basis of these results, since the v-*myc* oncogene is capable of inducing B-cell lymphomas in chicken, it was reasonable to speculate that the translocation of the c-*myc* oncogene to the heavy-chain locus may play a key role in the pathogenesis of Burkitt's lymphoma. This conclusion was also supported by the results of the analysis of Burkitt's lymphoma carrying the variant t(8;22) and t(2;8) chromosome translocations, since it was found that in these cases, the chromosome breakpoints involved a region immediately 3' (distal) to the involved c-*myc* locus (Figs. 1 and 2) (17,18).

Analysis of the human c-*myc* cDNA and comparison of the c-*myc* intron/exon boundaries indicated that the human c-*myc* oncogene has three exons; has a long, untranslocated leader sequence; and has the capability to code for a protein of 439 amino acids (19,20). Only the second and third exons code (Fig. 2). Analysis of the genomic c-*myc* DNA of the c-*myc* transcripts also showed that the c-*myc* oncogene has two promoters separated by approximately 160 nucleotides and that the

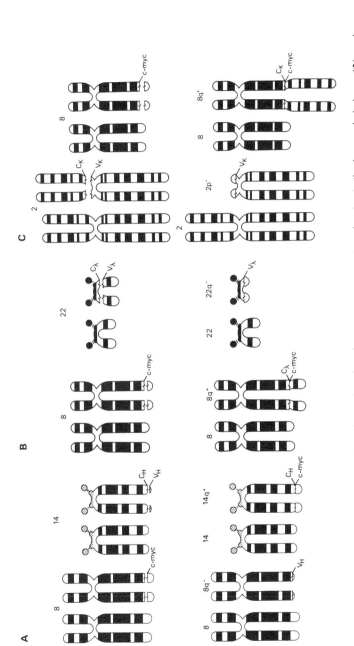

Fig. 1. In Burkitt's lymphoma with the t(8;14) translocation, the c-*myc* oncogene translocates to the heavy-chain locus (**A**), and a portion of the immunoglobuin locus (V_H) is translocated to chromosome 8. In Burkitt's lymphoma with the less frequent t(8;22) (**B**) and t(2;8) (**C**) translocations, the c-*myc* oncogene remains on the involved chromosome 8, but the genes for the immunoglobulin light-chain constant regions (C_κ and C_λ) translocate to a region 3' (distal) to the c-*myc* oncogene on the involved chromosome 8 ($8q^+$). Again, with these translocations, the immunoglobulin loci are split so that sequences that encode for the variable portion of the immunoglobulin molecule (V_κ or V_λ) remain on chromosome 2 or 22, respectively.

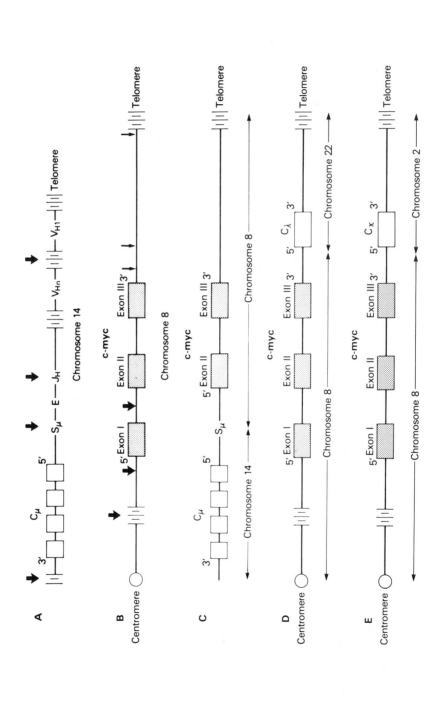

Fig. 2. A: A general diagram of an immunoglobulin heavy-chain gene. The constant region (C) contains DNA segments that encode the common portion of immunoglobulin molecules, the variable region (V_H) contains DNA segments that encode these portions of immunoglobulin molecules that differ from one another. During B-cell differentiation, immunoglobulin gene rearrangement occurs so that one V_H gene segment comes to lie immediately adjacent to a D segment (not shown), and the D segment lies adjacent to a short coding segment (J_H). The element labeled E is a DNA sequence that enhances promoter function, thereby increasing transcription of the productively rearranged immunoglobulin gene. The S region contains a DNA sequence that is involved in another type of DNA rearrangement that is the basis for the class switch from secretion of one type of heavy chain to another. The *vertical lines* indicate that distances between adjacent parts of the chromosome are not defined. In Burkitt's lymphoma with the t(8;14) translocation, the chromosome breakpoints within the heavy-chain locus may occur in the region carrying V_H genes; in the region between J_H and V_H in the heavy-chain joining segment (J_H); in the switch region (S_μ), or may involve a different constant-region coding segment. The *arrows* indicate possible sites for chromosomal breakpoints. **B**: A general diagram of the c-*myc* gene. In Burkitt's lymphoma with the t(8;14)I translocation, the chromosomal breakpoints on chromosome 8 are always 5' of the two coding exons (II and III) of the c-*myc* oncogene (*bold arrows*). In some cases, the c-*myc* oncogene is decapitated by the chromosomal break, and the first exon of the gene remains on the $8q^-$ chromosome, whereas the coding exons translocate to chromosome 14 (**C**). In Burkitt's lymphoma with the t(8;14) translocation with the less frequent t(8;22) and t(2;8) translocations, the breakpoints are distal to the c-*myc* oncogene (*this arrows*). **C**: Example of Burkitt's lymphoma with the t(8;14) translocation and a rearranged c-*myc* gene. The C_μ gene and the c-*myc* oncogene are inverted with respect to one another in that the transcriptional orientation of the c-*myc* and of the C_μ genes are in opposite directions (5'→3'). **D**: In Burkitt's lymphoma with the t(8;22) translocation, the c-*myc* oncogene remains on chromosome 8, whereas the portion of the λ locus that encodes the constant region of a light chain translocates to the chromosomal region 3' (distal) to the c-*myc* oncogene (**B**). **E**: In Burkitt's lymphomas with the t(2;8) translocation, the c-*myc* oncogene also remains on chromosome 8, whereas the constant region of the κ locus translocates to a chromosomal region 3' (distal) to the c-*myc* oncogene (**B**).

c-*myc* transcripts initiate from two different sites that initiate transcription (20). Since in the large majority of Burkitt's lymphoma the coding exons are intact, it is clear that alterations of the c-*myc* protein are not necessary in the development of Burkitt's lymphoma.

We have taken advantage of somatic cell hybrids between Burkitt's lymphoma cells and mouse myeloma cells to address the question of the genetic mechanisms involved in the pathogenesis of Burkitt's lymphoma. At first we attempted to determine whether there are differences in the expression of the translocated versus the normal c-*myc* oncogene in the hybrid cells (21). Since we found that hybrids containing the $14q^+$ chromosome expressed high levels of human c-*myc* transcripts, whereas hybrids containing the normal chromosome 8 did not express c-*myc* transcripts, we concluded that the translocated and the normal c-*myc* genes are under different regulatory control (21). This conclusion was supported by the analysis of *myc* transcripts in Burkitt's lymphoma cells carrying a rearranged c-*myc* gene (Fig. 2C), where we found the expression only of the translocated c-*myc* gene (22). It is interesting that when we examined somatic cell hybrids between mouse myeloma cells and human lymphoblastoid cells, we found that the human c-*myc* oncogene was repressed in hybrids (21). This result is consistent with the interpretation that although the normal c-*myc* oncogene can be downregulated in a plasma cell, the translocated c-*myc* gene fails to respond to the normal mechanisms of control of c-*myc* expression and is transcribed constitutively at elevated levels (21). Thus, the basic alteration in Burkitt's lymphoma with the t(8;14) chromosome translocation is that as a result of the close proximity of the involved c-*myc* oncogene to the heavy-chain locus, the c-*myc* oncogene is regulated like an immunoglobulin gene (21). This interpretation is supported by the fact that the expression of the translocated c-*myc* gene is B-cell specific (21,23). The results of the analysis of hybrids between mouse myeloma cells and Burkitt's lymphoma carrying the variant chromosome translocations are consistent with this interpretation (17,18), since the myeloma × Burkitt's lymphoma hybrids with the $8q^+$ chromosome (Fig. 2, D and E) express high levels of c-*myc* transcripts, whereas the hybrids containing the normal c-*myc* gene on normal chromosome 8 do not express c-*myc* transcripts (17,18).

It has been proposed that the deregulation of the involved c-*myc* oncogene is the result of changes in the first c-*myc* exon or of its decapitation (24). This interpretation is negated by the fact that we found a normal first exon in BL2 Burkitt's lymphoma with a t(8;22) chromosome translocation (17,48) and that a decapitated c-*myc* oncogene is not expressed in B cells in which a c-*myc* oncogene translocated to an immunoglobulin enhancer is expressed (25–27). Results of somatic cell hybridization and transfection experiments are consistent with the interpretation that the involved c-*myc* oncogene is deregulated because of its proximity to genetic elements present in the three immunoglobulin loci capable of activating gene transcription in *cis* over considerable chromosomal distances in B cells (25–27).

GENETIC ANALYSIS OF NON-BURKITT'S LYMPHOMA B-CELL NEOPLASMS

Specific reciprocal chromosomal translocations involving human chromosome 14 at band 14q32 are frequently observed in non-Burkitt's lymphoma B-cell malignancies (28,29). In the great majority of follicular lymphomas, the most common human B-cell malignancy, a translocation between chromosomes 14 and 18, t(14;18)(q32;q21) has been observed (28,30). A translocation between chromosomes 11 and 14, t(11;14)(q13;q32), has been observed in chronic lymphocytic leukemia of the B-cell type (31), in diffuse B-cell lymphoma, and in multiple myeloma (32). We have reasoned that since a specific oncogene, c-*myc*, is activated because of its proximity to the immunoglobulin loci in Burkitt's lymphoma, it should be possible to isolate putative oncogenes involved in B-cell malignancies carrying the t(14;18) and the t(11;14) by taking advantage of their proximity to the heavy-chain locus on chromosome 14 (33,34).

First, by fusing chronic lymphocytic leukemia (CLL) cells carrying a t(11;14) chromosome translocation with mouse myeloma cells, we found that the chromosome breakpoint in the leukemic cells directly involved the immunoglobulin heavy-chain locus (35). We then cloned the chromosome breakpoint involved in the t(11;14) translocation. By using probes flanking the breakpoints on chromosome 11 (33,36). Most of the t(11;14) chromosome translocations (Fig. 3) involve the joining (J_H) segment of the heavy-chain locus (36). Thus, we identified a locus, for which we proposed the name of bcl-1, that is activated by chromosomal translocation in B-cell malignancies (Fig. 4) (33,36). We have used the same approach to clone the segment of chromosome 18 that translocates to the heavy-chain locus in follicular lymphoma (34–37). We found that the involved locus, for which we proposed the name of bcl-2, is consistently rearranged in the great majority of follicular lymphomas (37). In most of the cases we examined, the bcl-2 gene (translocated to a J_H segment (Fig. 4) (37). Because of its translocation to the in-

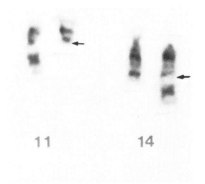

Fig. 3. The reciprocal t(11;14) (q13;q32) translocation in the neoplastic cells of a patient with diffuse large-cell lymphoma (LN87).

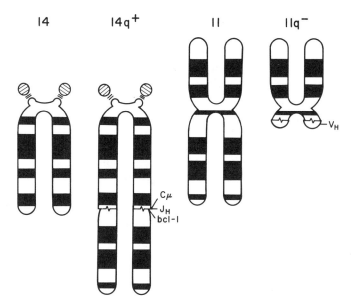

Fig. 4. Diagram of the t(11;14) (q13;q32) translocation observed in B-cell neoplasia. The breakpoint on chromosome 14 directly involves the H-chain locus. The V_H genes translocated to the deleted chromosome 11 (11q^-). The bcl-1 locus translocated from its normal position on band q13 of chromosome 11 to the involved chromosome 14.

volved heavy-chain gene, the bcl-2 gene is transcribed constitutively at elevated levels in the neoplastic B cells (Fig. 5) (37). Thus, this approach has permitted the isolation of two loci, bcl-1 and bcl-2, involved in the pathogenesis of human B-cell neoplasms.

The t(11;14) and t(14;18) Chromosome Translocations Involved in B-Cell Neoplasm Results from Mistakes in VDJ Joining

First, we cloned the chromosome breakpoints involved in the t(11;14) (q31;q32) chromosome translocations and the normal homologue of the bcl-1 locus on normal chromosome 11 (33). Southern blotting analysis of hybrid cell DNA using a DNA probe derived from the region flanking the J_H segment of heavy chain indicated that the hybridizing sequences are derived from human chromosome 11 (33). Sequence analysis of the breakpoints in two chronic lymphocytic leukemias of the B-cell type (CLL 1386 and CLL 271) carrying the t(11;14) chromosome translocation of the normal bcl-1 homologous sequences on chromosome 11 and of the involved J_H sequences on chromosome 14 is shown in Fig. 6. The two breakpoints on chromosome 11 are separated by eight nucleotides, whereas the breakpoints on chromosome 14 involve the J_4 segment of the heavy-chain locus (36). It is interest-

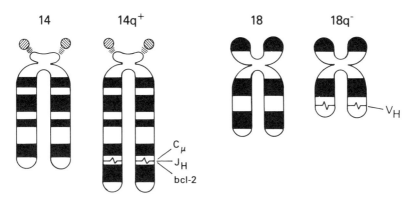

Fig. 5. Diagram of the t(14;18) translocation in human follicular lymphomas. The breakpoint on chromosome 14 directly involves the heavy-chain locus. The V_H genes are translocated to the involved chromosome 18 (18q⁻). The bcl-2 locus translocated from its normal position on band q21.3 of human chromosome 18 to the heavy-chain locus on chromosome 14.

ing to note that we found stretches of extranucleotides at joining sites (Fig. 6) (36). Since extranucleotides (N-regions) have also been shown to occur at joining sites in immunoglobulin and T-cell receptor genes (38 and since the breakpoints involved the 5' end of the involved J_H segment, we reasoned that the t(11;14) chromosome translocation may occur during the process of VDJ joining. It is very interesting that sequence analysis of the DNA of chromosome 11 flanking the breakpoint indicates that it contains signal sequences (7mer-9mer separated by a spacer of 12 nucleotides, underlined) for VDJ joining (Fig. 6). Thus, we conclude that the t(11;14)(q13;q32) chromosome translocations observed in B-cell neoplasms are the result of mistakes during the process of VDJ joining, where the VDJ joining enzyme joins separated segments of DNA from two different chromosomes, instead of joining two separated segments of DNA the same chromosome (36).

Analysis of the t(14;18) chromosome breakpoints involved in follicular lymphomas also indicated that the t(14;18) chromosome translocation involves J_H segments and occurs during the process of VDJ joining (39). Thus, the chromosome translocations involved in B-cell tumors are mediated by the enzymatic system involved in immunoglobulin gene rearrangement (36;39).

The Chromosome 14 Breakpoints in T-Cell Tumors Involve the Locus for the α-Chain of the T-Cell Receptor

We have cloned the human cDNA homologue of the murine gene specific for the α-chain of the T-cell receptor and used the human probe to map the chromosomal location of the human α-chain by Southern blotting analysis of mouse hu-

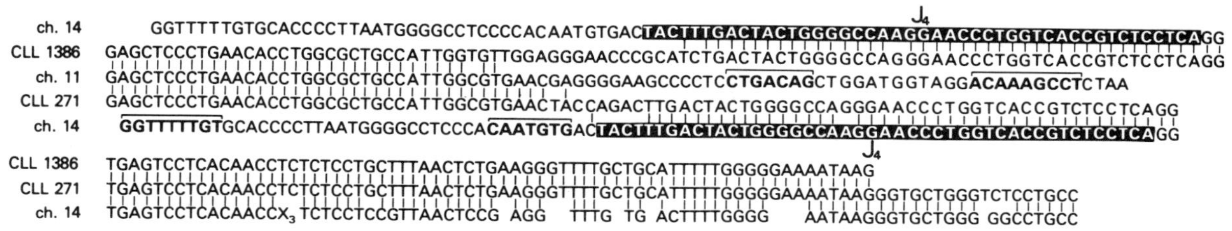

```
ch. 14                    GGTTTTTGTGCACCCCTTAATGGGGCCTCCCACAATGTGACTACTTTGACTACTGGGGCCAAGGAACCCTGGTCACCGTCTCCTCAGG
CLL 1386  GAGCTCCCTGAAACCTGGCGCTGCCATTGGTGTTGGAGGGAACCCGATCTGACTACTGGGGCCAGGGAACCCTGGTCACCGTCTCCTCAGG
ch. 11    GAGCTCCCTGAAACCTGGCGCTGCCATTGGCGTGAACGAGGGAAGCCCTCCTGACAGCTGGATGGTAGGACAAAGCCTCTAA
CLL 271   GAGCTCCCTGAAACCTGGCGCTGCCATTGGCGTGAACTACCAGACTTGACTACTGGGCCAGGGAACCCTGTCACCGTCTCCTCAGG
ch. 14    GGTTTTTGTGCACCCCTTAATGGGGCCTCCCACAATGTGACTACTTTGACTACTGGGGGCCAAGGAACCCTGGTCACCGTCTCCTCAGG
                                                                J₄
CLL 1386  TGAGTCCTCACAACCTCTCTCTGCTTTAACTCTGAAGGGTTTTGCTGCATTTTTGGGGGAAAATAAG
CLL 271   TGAGTCCTCACAACCTCTCTCTGCTTTAACTCTGAAGGGTTTTGCTGCATTTTTGGGGGAAAATAAGGGTGCTGGGTCTCCTGCC
ch. 14    TGAGTCCTCACAACCx₃TCTCCTCCGTTAACTCCGAGGTTTGTGACTTTTTGGGG AATAAGGGTGCTGGG GGCCTGCC
```

Fig. 6. DNA sequences of the joining sites between chromosomes 11 and 14 in CLL 271 and CLL 1386 and of corresponding normal chromosome 11. Identical nucleotide sequences are shown by vertical lines. The *boxed region* indicates the J₄ coding segment of the immunoglobulin heavy-chain gene. The DNA sequences shown by *brackets* on chromosome 14 indicate the conserved sequence 7mer-9mer (see text).

man hybrids and by *in situ* hybridization (39). We found that the gene for the α-chain of the T-cell receptor maps on chromosome 14 at band $q11.2$ (Fig. 7) (40). This result is of great interest, since region $14q11.2$ is frequently involved in rearrangements, translocations, and inversions in T-cell neoplasms (41–43).

In order to prove that the rearrangement directly involves the α-chain locus on chromosome 14, we have hybridized leukemic cells from the patients with T-cell acute lymphocytic leukemia (ALL) with a t(11;14) ($p13;q11$) with mouse T-cell leukemia cells (43). As shown in Table 1 and Fig. 8, analysis of the hybrids for markers of chromosome 11 and 14 indicates that the chromosome breakpoint splits the α-chain locus at band $14q11$ (Table 1; Fig. 8). Thus, the locus for the α-chain of the T-cell receptor is directly involved in the chromosomal rearrangements observed in T-cell tumors (44). These results strongly suggest that the genetic mechanisms involved in T-cell neoplasia are similar to those occurring in B-cell lymphomas and in other B-cell malignancies (45).

To confirm that this is really the case, we found two cases of T-cell malignancies in which the locus for the α-chain of the T-cell receptor is split by the chromosomal translocation and the α-gene is translocated to a region 3′ to the involved c-*myc* oncogene (Fig. 9) (46). The result of this translocation is a deregulation of

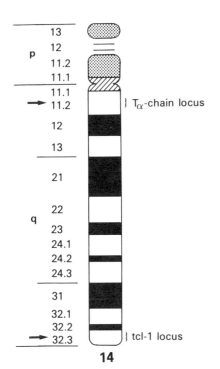

Fig. 7. Diagram of human chromosome 14. The *arrows* indicate the breakpoints observed in CLL cells with an inversion of chromosome 14. The *brackets* indicate the positions of the locus for the α-chain of the T-cell receptor (*upper*) and of the putative tcl-1 oncogene (*lower*). The chromosomal inversion and the t(14;14) ($q11;q32$) translocation should result in the juxtaposition of the Tα-chain locus and of the tcl-1 oncogene, thereby activating tcl-1.

TABLE 1. Presence of the V_α and C_α genes in hybrids between mouse BW5147 cells and human T ALL cells

Hybrids	Locus		Human markers [a]					Human chromosomes [b]			
	V_α	C_α	NP	LDH-A	c-H-ras	β-Globin	bcl-1	14	14q⁻	11	11p⁺
517 A-A3	+	+	+	+	+	+	−	++	+++	−	−
517 B-D3	+	+	+	+	+	+	++	−	+	−	++
517 B-B1	−	−	−	+	+	+	++	−	+	−	++
517 B-A3	+	−	+	+	+	+	−	−	+	−	−
517 B-D3-G8	+	−	+	+	+	+	−	−	++	−	−
517 B-D3-G9	+	−	+	+	+	+	+	−	++	−	+++
517 B-D3-D2A	−	+	−	−	−	−	+	−	−	−	++
517 A-A3-A10	+	+	+	+	+	+	−	+	++	−	−
517 A-A3-G7	+	−	+	+	+	+	−	−	+++	−	−
515 BD2-CF3	+	−	+	+	+	+	−	−	+	−	−
515 BD2-CF6	+	−	+	+	+	+	−	−	+++	−	−

[a](NP) nucleoside phosphorylase; (LDH-A) lactic dehydrogenase A.
[b]Frequency of metaphases with relevant chromosomes; (−) none; (+) 10–30%; (++) 30–50% (+++) > 50%. At least 25 metaphases were examined for each hybrid after trykpsin-Giemsa staining. Selected metaphases were studied by the G11 technique to confirm the human origin or relevant chromosomes.

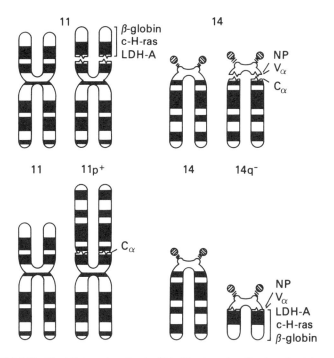

Fig. 8. The t(11;14)(q13;q11) translocation in ALL. The translocation breakpoint on chromosome 14 splits the locus for the α-chain of the T-cell receptor. The V_α genes remain on the $14q^-$ chromosome, whereas the C_α genes translocate to the involved chromosome 11 ($11p^+$). The gene for human nucleoside phosphorylase remains on the involved chromosome 14 ($14q^-$). The genes LDH-A, β-globin, and c-H-*ras* translocate to the involved chromosome 14 ($14q^-$).

the c-*myc* involved in the translocation that is transcribed constitutively at elevated levels (46). Thus, the molecular mechanisms involved in T-cell tumors parallel those involved in the pathogenesis of B-cell malignancies (45). Although the T-cell leukemias we have studied, which have the t(11;14) and the t(8;14) chromosome translocations, are not related to the human T-cell leukemia virus (HTLV-1), Sandamori et al. (47) indicate that there are abnormalities (translocations and inversions) of chromosome 11 at band 14q11 in Japanese patients with T-cell leukemias. This suggests that HTLV-1 may not be leukemogenic per se. This virus may play an important role in leukemogenesis, perhaps analogous to the role of Epstein-Barr virus in African Burkitt's lymphoma, by expanding the population of lymphocytes at risk of developing specific chromosome translocations. Thus, activation of proto-oncogenes in T-cell neoplasms, because of their proximity to the locus for the α-chain of the T-cell receptor, may be the critical step in the pathogenesis of many T-cell malignancies.

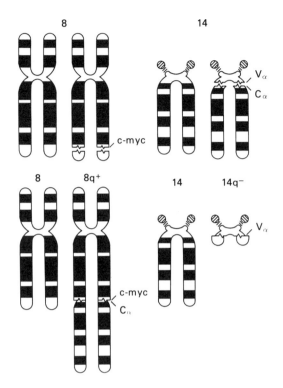

Fig. 9. Diagram of the t(8;14) (24, q11) chromosome translocation in T-cell leukemias. The breakpoint on chromosome 8 is distal to the involved c-*myc* oncogene, whereas the breakpoint on chromosome 14 directly involves the locus for the α-chain of the T-cell receptor.

CONCLUSIONS

The data presented in this review indicate that chromosomal translocations, at least in B-cell tumors and probably in T-cell tumors, are catalized by the same enzymatic system that is involved in immunoglobulin and T-cell receptor gene rearrangement, respectively (36,39). Although the translocations involve the mechanism of VDJ joining in most of B-cell tumors with translocations involving 14q32, it is still not clear which molecular events are involved in the t(8;14) chromosomal translocations of sporadic cases of Burkitt's lymphoma where one of the switch regions is involved. It seems likely, however, that these rearrangements may occur during the process of human immunoglobulin isotype switching.

The result of the translocations is a deregulation of the oncogene or putative oncogene flanking the chromosomal translocation breakpoints, because of its close proximity to genetic elements within the immunoglobulin or T-cell receptor loci capable of activating gene transcription *in cis* over considerable chromosomal distances. Thus, lymphomas and leukemia cells carrying specific chromosomal translocations or inversions can be utilized to clone and characterize genes involved in the pathogenesis of human hematopoietic malignancies (45). Therefore, these specific chromosome rearrangements provide us with the opportunity not only to define the molecular genetic events leading to neoplasia, but also to isolate human genes involved in the control of cell proliferation and in the pathogenesis of human cancer.

ACKNOWLEDGMENT

This work was supported by grants from the National Institutes of Health, CA 16685, CA 25875, CA 39860 (C.M.C.), and CA 35150 (P.C.N.).

REFERENCES

1. Yunis, J.J., Oken, M.M., Kaplan, M.E., Ensrud, K.M., Howe, R.R., and Theologides, A. (1982): Distinctive chromosomal abnormalities in histologic subtypes of non-Hodgkin's lymphoma. *N. Engl. J. Med.*, 307:1231–1236.
2. Nowell, P.C., Hungerford, D.A. (1960): A minute chromosome in chronic granulocytic leukemia. *Science*, 132:1497.
3. Dalla Favera, R., Bregni, M., Erikson, J., Patterson, D., Gallo, R.C., and Croce, C.M. (1982) Assignment of the human c-*myc* oncogene to the region of chromosome 8 which is translocated in Burkitt lymphoma cells. *Proc. Natl. Acad. Sci. U.S.A.*, 79:7824–7825.
4. Groffen, J., Stephenson, J.R., Heisterkamp, N., deKlein, A., Bartram, C.R., and Grosveld, G. (1984): Philadelphia chromosomal breakpoints are clustered within a limited region, bcr, on chromosome 22. *Cell*, 36:93–99.
5. Manolov, G. and Manolova, Y. (1972): Marker band in one chromosome 14 from Burkitt lymphoma. *Nature*, 237:33–34.
6. Zech, L., Haglund, V., Nilson, N., and Klein, G. (1976): Characteristic chromosomal abnormalities in biopsies and lymphoid cell lines from patients with Burkitt and non-Burkitt lymphomas. *Int. J. Cancer*, 17:47–56.
7. van den Berghe, H., Parloir, C., Gosseye, S., Eglebienne, V., Cornu, G., and Sokal, G. (1979): *Cancer Genet. Cytogenet.*, 1:9–14.
8. Lenoir, G.M., Preud'Homme, J.L., Bernheim, A., and Berger, R. (1982): Correlation between immunoglobulin light chain expression and variant translocation in Burkitt's lymphoma. *Nature*, 298:474–476.
9. Croce, C.M., Shander, M., Martinis, J., Cicurel, L., D'Ancona, G.G., Dolby, T.W., and Koprowski, H. (1979): Chromosomal location of the human immunoglobulin heavy chain genes. *Proc. Natl. Acad. Sci. U.S.A.*, 76:3416–3419.
10. Erikson, J., Martinis, J., and Croce, C.M. (1981): Assignment of the human genes for lambda immunoglobulin chains to chromosome 22. *Nature*, 294:173–175.
11. Klein, G. (1981): The role of gene dosage and genetic transposition in carcinogenesis. *Nature*, 294:313–318.
12. McBride, D.W., Heiter, P.A., Hollis, G.F., Swan, D., Otey, M.C., and Leder, P. (1982): Chromosomal location of human kappa and lambda immunoglobulin light chain constant region genes. *J. Exp. Med.*, 155:1480–1490.
13. Malcolm, S., Barton, P., Murphy, C., Ferguson-Smith, M.A., Bentley, D.L., and Rabbitts, T.H. (1982): *Proc. Natl. Acad. Sci. U.S.A.*, 79:4957–4961.
14. Erikson, J., Finan, J., Nowell, P.C., and Croce, C.M. (1982): Translocation of immunoglobulin V_H genes in Burkitt lymphoma. *Proc. Natl. Acad. Sci. U.S.A.*, 80:810–824.
15. Dalla Favera, R., Martinotti, S., Gallo, R.C., Erikson, J., and Croce, C.M. (1983): Translocation and rearrangements of the c-*myc* oncogene in human differentiated B cell lymphomas. *Science*, 219:963–967.
16. Adams, J.M., Gerondakis, S., Webb, E., Corcoran, C.M., and Cory, S. (1983): Cellular myc oncogene is altered by chromosome translocation to an immunoglobulin locus in murine plasmacytomas and is rearranged similarly in human Burkitt lymphomas. *Proc. Natl. Acad. Sci. U.S.A.*, 80:1982–1986.

17. Croce, C.M., Theirfelder, W., Erikson, J., Nishikura, K., Finan, J., Lenoir, G., and Nowell, P.C. (1983): Transcriptional activation of an unrearranged and untranslocated c-*myc* oncogene by translocation of a C lambda locus in Burkitt lymphoma. *Proc. Natl. Acad. Sci. U.S.A.*, 80:6922–6926.
18. Erikson, J., Nishikura, K., ar-Rushdi, A., Finan, J., Emanuel, B., Lenoir, G., Nowell, P.C., and Croce, M. (1983): Translocation of a kappa immunologuin locus to a region 3' of an unrearranged c-*myc* oncogene enhances c-*myc* transcription. *Proc. Natl. Acad. Sci. U.S.A.*, 80:7581–7885.
19. Watt, R., Stanton, L.W., Marcu, K.B., Gallo, R.C., Croce, C.M., and Rovera, G. (1983): Nucleotide sequence of cloned cDNA of the human c-*myc* gene. *Nature*, 303:725–728.
20. Watt, R., Nishikura, K., Sorrentino, J., ar-Rushdi, A., Croce, C.M. and Rovera, G. (1983): The structure and nucleotide sequence of the 5' end of the human c-*myc* gene. *Proc. Natl. Acad. Sci. U.S.A.*, 80:6307–6311.
21. Nishikura, K., ar-Rushdi, A., Erikson, J., Watt, R., Rolvera, G., and Croce, C.M. (1983): Differential expression of the normal and of the translocated human c-*myc* oncogene in B cells. *Proc. Natl. Acad. Sci. U.S.A.*, 80:4822–4826.
22. ar-Rushdi, A., Nishikura, K., Erikson, J., Watt, R., Rovera, G., and Croce, C.M. (1983): Differential expression of the translocated and of the untranslocated c-*myc* gene in Burkitt lymphoma. *Science*, 222:390–393.
23. Nishikura, K., ar-Rushdi, A., Erikso, J., DeJesus, E., Dugan, D., and Croce, C.M. (1984): Repression of rearranged u gene and translocated c-*myc* in mouse 3T3 cells × Burkitt lymphoma cell hybrids. *Science*, 224:399–402.
24. Leder, P., Battey, J., Lenoir, G., Maulding, G., Murphy, W., Potter, H., Stewart, T., and Taub, R. (1983): Translocations among antibody genes in human cancer. *Science*, 222:765–767.
25. Croce, C.M., Erikson, J., ar-Rushdi, A., Aden, D., and Nishikura, K. (1984): The translocated c-*myc* oncogene of Burkitt lymphoma is transcribed in plasma cells and repressed in lymphoblastoid cells. *Proc. Natl. Acad. Sci. U.S.A.*, 81:3170–3174.
26. Croce, C.M., Erikson, J., Huebner, K., and Nishikura, K. (1985): Co-expression of the translocated and of the normal c-*myc* oncogene in hybrids between Daudi and lymphoblastoid cells. *Science*, 227:1235–1238.
27. Feo, S., Harvey, R., Showe, K., and Croce, C.M. (1986): Regulation of the translocated c-*myc* genes transfected into plasmacytoma cells. *Proc. Natl. Acad. Sci. U.S.A.*, (in press).
28. Yunis, J. (1983): The chromosomal basis of human neoplasia. *Science*, 221:227–236.
29. Rowley, J.D. (1983): Identification of the constant chromosome regions involved in human hematologic malignant disease. *Science*, 216:749–751.
30. Fukuhara, S., Rowley, J.D., Variakojis, D., and Golomb, H.M. (1979): Chromosome abnormalities in poorly differentiated lymphocytic lymphoma. *Cancer Res.*, 39:3119–3128.
31. Nowell, P.C., Shankey, T.V., Finan, J., Guerry, D., and Besa, E. (1981): Proliferation, differentiation and cytogenetics of chronic leukemic B lymphocytes cultured with mitomycin-treated normal cells. *Blood*, 57:444–451.
32. van den Berghe, H., Vermaelen, K., Louwagie, A., Criel, A., Mecucci, C., and Vaerman, J.P. (1984): High incidence of chromosome abnormalities in IgG3 myeloma. *Cancer Genet. Cytogenet.*, 11:381–387.
33. Tsujimoto, Y., Yunis, J., Onorato-Showe, L., Erikson, J., Nowell, P.C., and Croce, C.M. (1984): Molecular cloning of the chromosomal breakpoint of 8-cell lymphomas and leukemias with the t(11;14) chromosome translocation. *Science*, 224:1403–1406.
34. Tsujimoto, Y., Finger, L.R., Yunis, J., Nowell, P.C., and Croce, C.M. (1984): Cloning of the chromosome breakpoint in neoplastic B cells with the t(14;18) chromosome translocation. 226:1097–1099.
35. Erikson, J., Finan, J., Tsujimoto, Y., Nowell, P.C., and Croce, C.M. (1984): The chromosome 14 breakpoint in neoplastic B cells with the t(11;14) translocation involves the immunogloblin heavy chain locus. *Proc. Natl. Acad. Sci. U.S.A.*, 81:4144–4148.
36. Tsujimoto, Y., Jaffe, E., Cossman, J., Gorham, J., Nowell, P.C., and Croce, C.M. (1985): Clustering of breakpoints on chromosome 11 in human B cell neoplasms with the t(11;14) chromosome translocation. *Nature*, 315:340–343.
37. Tsujimoto, Y., Cossman, J., Jaffe, E., and Croce, C.M. (1985): Involvement of the bcl-2 gene in human follicular lymphoma *Science*, 228:1440–1443.

38. Desidero, S.V., Yancopoulos, D.G., Paskind, M., Thomas, E., Boss, M.A., Landau, N., Alt, F.W., and Baltimore, D. (1984): Insertion of N regions into heavy-chain genes is correlated with expression of terminal deoxytransferase in B cells. *Nature*, 311:752–755.
39. Tsujimoto, J., Gorham, J., Cossman, J., Jaffe, E., and Croce, C.M. (1985): The t(14;18) chromosome translocations involved in B cell neoplasms results from mistakes in VDJ joining. *Science*, 229:1390–1393.
40. Croce, C.M., Isobe, M., Palumbo, A., Puck, J., Ming, J., Tweardy, D., Erikson, J., Davis, M., and Rovera, G. (1985): The gene for the alpha-chain of human T cell recpetor: Location on chromosome 14 region involved in T cell neoplasms. *Science*, 227:1044–1047.
41. Zech, K., Gahrton, G., Hammerarstrom, L., Juliusson, G., Mellstedt, H., Robert, K.H., and Smith, C.I.E. (1984): Inversion of chromosome 14 marks human T-cell chronic lymphocytic leukemia. *Nature*, 308:858–860.
42. Hecht, F., Morgan, R., Kaiser-McCaw Hecht, B., and Smith, S.D. (1984): Common region on chromosome 14 in T-cell leukemia and lymphoma. *Science*, 226:1445–1446.
43. Williams, D.L., Look, A.T., Melvin, S.L., Roberson, P.K., Dahl, G. Flake, T. and Stass, S. (1984): New chromosomal translocations correlate with specific immunophenotypes of childhood acute lymphoblastoid leukemia. *Cell*, 36:101–109.
44. Erikson, J., Williams, D.L., Finan, J., Nowell, P.C., and Croce, C.M. (1985): Locus of the alpha-chain of the T-cell receptor is split by chromosome translocation in T-cell leukemias. *Science*, 229:784–786.
45. Croce, C.M., and Nowell, P.C. (1985): Molecular Basis of Human B Cell Neoplasia. *Blood*, 65:1–7.
46. Erikson, J., Finger, L., Sun, L., Minowada, J., Finan, J., Emanuel, B.S., Nowell, P.C., and Croce, C.M. (1986): c-*myc* deregulation by translocation of the locus for the alpha-chain of the T-cell leukemias. *Science*, 232:884–886.
47. Sandamori, N., Kusana, M., Nishino, K., Tagawa, M., Yao, E., Yamada, Y., Amagasaki, T., Kinoshita, K., and Ichimuru, M. (1985): Abnormalities of chromosome 14 at band 14q11 in Japanese patients with adult T-cell leukemia. *Cancer Genet. Cytogenet.*, 17:279–282.
48. Erikson, J., and C.M. Croce (1986): *In preparation.*

Multistep Cytogenetic Scenario in Chronic Myeloid Leukemia

Sverre Heim and Felix Mitelman

Department of Clinical Genetics, University Hospital, S-221 85 Lund, Sweden

The description more than a quarter of a century ago by Nowell and Hungerford (42) of an unusually small G-group chromosome in leukemic cells from patients with chronic myeloid leukemia (CML) marked the beginning of a new era in human oncogenetics. For the first time, a consistent chromosome abnormality had been described in a human malignancy. The newly detected chromosomal marker was named the Philadelphia chromosome, in honor of the city in which it was discovered, and abbreviated Ph^1, emphasizing that this was the first cancer-specific cytogenetic marker to be reported from that city. The whole system of marker nomenclature was subsequently changed, but for sentimental reasons the abbreviation Ph^1 remained. Initially, no method was available to determine which small, acrocentric chromosome gave rise to the Ph^1 marker; neither was it known whether a deletion or a translocation was the mechanism underlying the abnormality. However, the general sentiment at the time was that Ph^1 was probably derived from a chromosome of pair 21, and a simple deletion with loss of genetic material seemed the most straightforward mechanism to explain its origin, particularly since no cytogenetic evidence of any translocation was detectable in the preparations then available (55).

Intensive investigations during the following decade culminated in the early 1970s, when improved cytogenetic techniques, especially the introduction of chromosome banding methods, made it possible to prove unequivocally that Ph^1 was in fact a deranged chromosome 22 (4,43,48). Finally, in 1973, analysis of banded preparations clearly demonstrated that the Ph^1 chromosome in most cases originated not through a simple deletion of chromosome 22, but through a translocation of chromosomal material from chromosome 22 to the terminal part of the long arm of chromosome 9, i.e., t(9;22)($q34;q11$) (49). However, even after the introduction of banding techniques, the power of resolution in cytogenetics has been insufficient to establish whether or not the translocation is actually reciprocal, i.e., whether genetic material is also moved from chromosome 9 to the deranged chromosome 22. Such reciprocity was conclusively demonstrated in 1982 by de Klein et al. (6) with the help of *in situ* hybridization. Thus, consistent improvements in investigative techniques have for more than two decades enabled a constant expansion of our knowledge about the Ph[1] chromosome.

Concomitant with these efforts, cytogenetic research in literally thousands of cases of human leukemia has established beyond doubt that not only is the Ph[1] marker a very consistent finding in CML [approximately 85% of patients with typical CML have the marker (55)], but Ph[1] may also be found in both acute lymphocytic leukemia (ALL) and, more rarely, in acute nonlymphocytic leukemia (ANLL) (11,60). Furthermore, in 5% to 10% of patients with Ph[1]-positive CML, Ph[1] originates through a mechanism other than the classical t(9;22) (17,37,44,54). A comparable number of patients with chronic myeloproliferative disease clinically and hematologically indistinguishable from CML have no Ph[1] chromosome (37,55). To complicate the picture further, with progression of CML through accelerated to blastic phase, a number of additional, secondary chromosomal changes may accrue, making the karyotype more complex as the disease becomes more malignant.

Before quantitatively addressing the question of the multistep evolution of chromosome abnormalities in CML, it might be useful first to summarize the nomenclature used when describing chromosomal phenomena and then discuss the exact nature of t(9;22), the cytogenetic hallmark of the great majority of CML cases.

CYTOGENETIC NOMENCLATURE

The nomenclature of chromosome description has been standardized at a series of conferences, the first of which was held in 1960 and the most recent in 1977 (23). In the nomenclature now recommended, each chromosome is seen as a continuous series of transverse bands with no unbanded areas so that by definition, no "interbands" exist. The banding pattern along each chromosome may be produced by various staining methods. The bands are allocated to certain defined regions along the chromosome arms, and the regions are delimited by specific landmarks; a region is defined as the area between two adjacent landmarks.

Regions and bands are numbered consecutively from the centromere outward along each chromosome arm. The symbols p and q designate the short and long arm of each chromosome, respectively. In designating a particular band, four items are required: the chromosome number, the arm symbol, the region number, and the band number within that region. These items are given in that order and without spacing or punctuation; for example, $9q34$ indicates chromosome 9, long arm, region 3, band 4.

The location of any given breakpoint is specified by the band in which the break has occurred. Structurally altered chromosomes may thus be defined by their breakpoints. These are specified within parenthesis immediately following the description of the type of rearrangement and the chromosome(s) involved; for example, translocations are specified by the symbol "t," deletions by "del," duplications by "dup," inversions by "inv," isochromosomes by "i," insertions by "ins," and dicentric and ring chromosomes by "dic" and "r," respectively. Hence, t(9;22)(q34;q11) indicates a translocation between chromosomes 9 and 22, the breakpoint in chromosome 9 is localized to the long arm, region 3, band 4; the breakpoint in chromosome 22 is in the long arm, region 1, band 1. A marker (mar) or a derivative (der) chromosome is a structurally abnormal chromosome, which, if the banding pattern is recognized, may be adequately described by the standard nomenclature outlined above.

Plus (+) and minus (−) signs are placed before the appropriate symbol to indicate additional or missing whole chromosomes. They may be placed after the symbol to indicate an increase or decrease in the length of a chromosome, a chromosomal arm, or region.

A clone is a population of cells derived from a single progenitor. In describing acquired chromosomal aberrations, it is reasonable and common practice to infer a clonal origin when a number of cells have the same or related abnormal chromosomal complement. Thus, a clone is not necessarily homogeneous. It will always contain at least two cells with the same aberration; but, in some cases, especially with hypodiploid cells, three or even more cells may be required to determine whether the abnormality is clonal, all depending on the number of cells analyzed.

CYTOGENETIC MORPHOLOGY OF THE (9;22) TRANSLOCATION

A typical metaphase from a patient with CML, with the standard t(9;22) as the only abnormality is shown in Fig. 1. Whereas the consensus has always been that the breakpoint on chromosome 9 is in $q34$, some controversy has existed regarding the breakpoint on chromosome 22. The great majority of reports have assigned the breakpoint to $22q11$, but even the more distal bands $q12$ and $q13$ have been implicated. Prakash and Yunis (47) studied the morphology of the Ph[1]-producing translocations in detail using synchronizing techniques to increase the yield of mitoses in prophase and early prometaphase (62). This high-resolution investigation included Ph[1]-positive patients with ALL and ANLL in addition to the typical CML patients, and concluded that the translocations seemed identical at the resolution attained (400–1000 bands per haploid set, as opposed to the standard 300 bands per karyotype obtainable with conventional midmetaphase chromosomes). The breakpoint on chromosome 9 was consistently mapped to subband $9q34.1$ and the breakpoint on chromosome 22 to subband $22q11.2$. A schematic representation of chromosome pairs 9 and 22 exemplifying the subband resolution that may be achieved in late prophase is provided in Fig. 2.

A word of caution may be appropriate with regard to the concept of "identical breakpoints" in the cytogenetic meaning of the words. What seems like indistinguishable breakage sites have in actual fact with molecular genetic methods been demonstrated to be several kilobases (kb) apart in the few patients yet studied in this manner (13,20,57), both on $22q$ and even more on $9q$. Hence, the uniformity of the translocations may not be as massive as one might surmise on the basis of

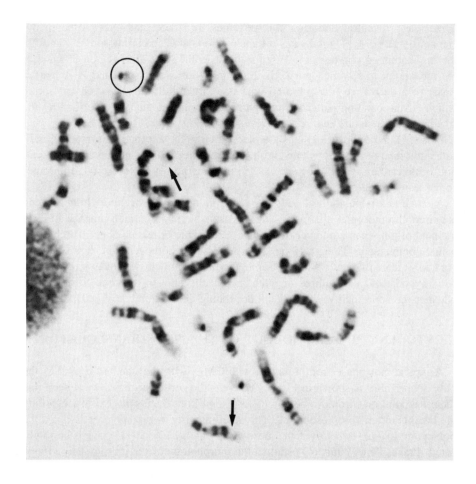

Fig. 1. Typical metaphase from CML patient with t(9;22)(q34;q11) as the only abnormality. The *arrows* indicate the breakpoints in the two rearranged chromosomes; the *circle* surrounds the normal chromosome 22.

cytogenetic investigations alone. Whether further detailed studies of t(9;22), preferably combining cytogenetic and molecular genetic techniques, may again increase the specificity of the Ph[1]–CML association, perhaps by revealing systematic differences among the translocations in Ph[1]-positive CML, ALL, and ANLL, is an intriguing but still unanswered question.

CLASSICAL Ph[1]-POSITIVE CML—DISTRIBUTION OF SECONDARY CHANGES

The typical clinical course for classical t(9;22)-associated CML is to proceed through a chronic, relatively quiescent phase of a few years' duration before enter-

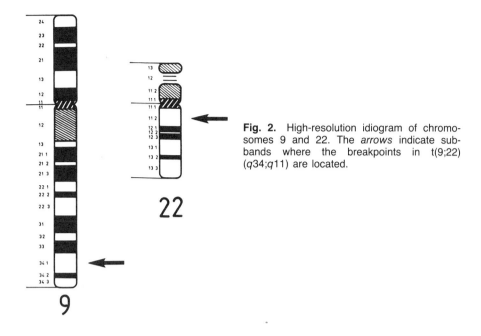

Fig. 2. High-resolution idiogram of chromosomes 9 and 22. The *arrows* indicate subbands where the breakpoints in t(9;22) (*q*34;*q*11) are located.

ing an accelerated, more malignant stage. Finally, the disease reaches the blast phase, which may be indistinguishable from ANLL.

Characteristically, during the chronic phase, t(9;22) is the only detectable cytogenetic abnormality, although additional chromosomal changes may be present early in the disease (53). The malignant cells are clonal in origin, as demonstrated in studies using enzyme markers (7) and fluorescent chromosome polymorphisms (12,16,30). Approximately 75% to 80% of chronic phase CML patients develop additional aberrations when entering the acute, blastic phase (2,10,30,39,51,54,55). Such secondary changes, which have even been detected in extramedullary tissues (41), may precede the hematologic and clinical manifestations of more active disease by several months, and hence serve as valuable prognostic indicators (10).

Early investigations before banding techniques were available indicated that the chromosomal changes occurring in excess of Ph^1 in CML were nonrandom (5,45). This has now been conclusively confirmed by several investigators (e.g., 2,10,14,30,37,39,51,53). Mitelman et al. (39) coined the terms major and minor routes for the pathways of cytogenetic evolution followed by CML cells in the blast phase. Major route changes were the acquisition of an extra Ph^1 chromosome, an extra chromosome 8, and/or the occurrence of i(17q).

In an attempt to ascertain the present state of knowledge with regard to additional chromosome aberrations in CML patients with t(9;22) as the primary abnormality, we have surveyed the data in our computer-based tumor karyotype registry (34), which comprise cytogenetic data collected from three sources: (a) published cases, ascertained with the aid of three separate computer-based literature scans; (b) unpublished cases kindly communicated by numerous colleagues; and (c) unpublished cases from our laboratory. By January 1985, 5,345 human malignancies were included in the data base.

No less than 709 CML patients with t(9;22) and additional chromosome abnormalities were retrieved. A total of 2,005 aberrations (Table 1) were found among the patients (reciprocal translocations are entered in the table as two aberration events, one for each chromosome affected). The main conclusions with regard to secondary karyotype evolution reached in previous reports remain valid in this more comprehensive survey. Clearly, chromosomes 8, 17, and 22 are the chromosomes most often involved in secondary changes. This preferential engagement is largely due to the frequent occurrence of trisomy 8, i(17q), and an extra Ph^1. The somewhat less prominent excess of trisomy 19 is also striking. Finally, gain of an extra chromosome 21 is the fifth most common additional aberration, occurring slightly more often than the loss of a sex chromosome. A graphic presentation of the information presented in Table 1 is provided in Fig. 3.

With on average almost three secondary changes per patient (709 patients share the 2,005 aberrations), it is obvious that combinations of several additional aberrations must be present in many cases. Investigations based on more limited numbers have pointed to the preferential occurrence of some combinations of aberrations in blast phase CML (2,39,55), e.g., the simultaneous finding of +8 and i(17q) or +8 and $+Ph^1$, whereas other secondary combinations, such as $+Ph^1$ and i(17q) or i(17q) and +19, were almost nonexisting. In order to ascertain how frequently the different permutations of the various common changes in CML occur, we have reviewed the data registry (34), looking specifically for all the characteristic secondary changes and combinations of changes:

+8; i(17q); $+Ph^1$; +19; +8, $+Ph^1$; +8, +19; +8, i(17q);

i(17q), $+Ph^1$; i(17q), +19; +8, i(17q), $+Ph^1$; +8, i(17q), +19;

+8, $+Ph^1$, +19; i(17q), $+Ph^1$, +19; +8, $+Ph^1$, i(17q), +19

The information retrieved is presented in condensed form in Table 2. Cases containing none of these secondary abnormalities were referred to the "other aberrations" category. There is good general agreement between the distribution of frequencies in Table 2 and corresponding data based on smaller patient samples (39,51). Gain of chromosome 8, an extra Ph^1, or i(17q) are the main additional changes, occurring with comparable frequencies (40%, 38%, and 28%, respectively). Gain of chromosome 19, on the other hand, seems to occur later in karyotype evolution, most often in combination with both +8 and $+Ph^1$. Furthermore, whereas i(17q) alone or in combination with +8 is a quite frequent phenomenon, the combinations i(17q), +19; +8, i(17q), +19; and i(17q), $+Ph^1$, +19 are only very rarely seen. The combination $+Ph^1$, i(17q) is also a rare finding, further testifying to the apparently restrictive role of i(17q) in the cytogenetic evolution in CML, at least when no extra chromosome 8 is present in the cells.

TABLE 1. Secondary chromosome aberrations in 709 patients with classical t(9;22) associated CML[a]

Chromosome	+	−	t	del	inv	i	ins	der	dic	dup	r	Total	%
1	7	6	36	8	1	1	1			3		63	3.1
2	3	5	12	4								24	1.2
3	9	8	20	6	1	1						46	2.3
4	8	8	11	1								28	1.4
5	12	11	14	6								43	2.1
6	28	7	9	12	2							56	2.8
7	10	22	18	3				1			2	58	2.9
8	307	13	15	4		1		1				340	16.9
9	21	10	14	2				1		1		50	2.5
10	31	9	5	2		1						47	2.3
11	14	9	9	3	2		1				2	40	2.0
12	21	6	13	1								41	2.0
13	17	7	13	7		1						45	2.2
14	18	7	12									37	1.8
15	12	11	6	6								35	1.7
16	8	14	8	3								33	1.6
17	48	23	36	2		192					1	302	15.0
18	12	11	5								2	30	1.5
19	110	12	6	2							1	131	6.5
20	8	11	11	1								31	1.5
21	60	5	13									78	3.9
22	11	12	4	270		2		1	10			310	15.4
X	5	8	5	1		1						20	1.0
Y	8	48		2	1							59	2.9
mar	60											60	3.0
dmin	2											2	0.1
Total:	850	283	295	346	7	199	2	5	10	4	8	2,005	

[a] (+) gain of whole chromosome; (−) loss of whole chromosome; (t) translocation; (del) deletion; (inv) inversion; (i) isochromosome; (ins) insertion; (der) derivative chromosome; (dic) dicentric chromosome; (dup) duplication; (r) ring chromosome; (mar) marker chromosome; (dmin) double minutes.

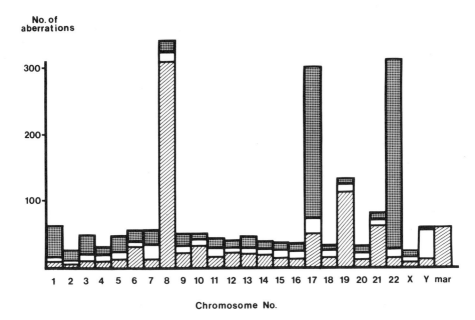

Fig. 3. Graphic presentation of the distribution of secondary chromosome aberrations in classical Ph¹-positive CML: (*hatched area*) gain of whole chromosomes; (*open area*) loss of whole chromosomes; (*dotted area*) structural rearrangements.

TABLE 2. *Frequency of various combinations of secondary chromosomal changes in 709 patients with classical Ph¹-positive CML*

Combination	No. of patients	Relative frequency %
+8	57	8.0
i(17q)	83	11.7
+Ph¹	97	13.7
+19	7	1.0
+8, +Ph¹	60	8.5
+8, +19	14	2.0
+8, i(17q)	63	8.9
i(17q), +Ph¹	5	0.7
i(17q), +19	1	0.1
+Ph¹, +19	22	3.1
+8, i(17q), +Ph¹	23	3.2
+8, i(17q), +19	6	0.8
+8, +Ph¹, +19	47	6.6
i(17q), +Ph¹, +19	0	0.0
+8, +Ph¹, i(17q), +19	14	2.0
Other aberrations	210	29.6

A graphic summary of the secondary changes ascertained is presented in Fig. 4. The outline of the graph is based on a similar diagram presented previously (39) and attempts to emphasize the dynamic aspects of the karyotypic evolution. The "minor routes" category covers all rearrangements grouped as "other aberrations" in Table 2, including 32 patients with loss of one sex chromosome as their sole additional change. The development through major pathways involving +8, i(17q), and +Ph^1 toward the more complex karyotypic combinations is a clear-cut example of the nonrandomness of chromosomal change even in the later stages of this disease.

Bearing in mind that t(9;22) is a relatively frequent primary abnormality even in the acute leukemias, it is of course of interest to find out what similarities or

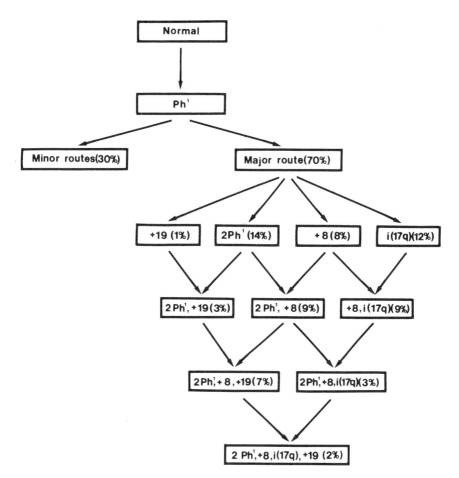

Fig. 4. Quantitative assessment of major route secondary chromosome aberrations in classical Ph^1-positive CML.

differences exist in the distribution of secondary changes in the chronic and acute disease varieties. An investigation including 60 ANLL cases and 61 ALL cases with t(9;22) as the primary abnormality, all ascertained from the same data bank (34), has been undertaken with the aim of determining the secondary changes occurring during disease progression (18). No clear group differences were detected between these ALL and ANLL patients with regard to the distribution of secondary changes. Chromosome 7 involvement, in particular monosomy 7, was the most prominent secondary rearrangement detected in both t(9;22)-associated acute leukemia groups. Trisomy 8 or $+Ph^1$ was somewhat less frequent but still relatively characteristic features; i(17q) did occur but not more often than several other additional abnormalities. Hence, a difference seems to exist between the distribution of secondary chromosome changes in chronic and acute Ph^1-positive leukemias. Whereas $+8$ and $+Ph^1$ are prominent features in both groups, chromosome 7 rearrangements (especially monosomy) are much less common in CML. It may therefore well be that nonchromosomal factors characteristic of the acute and chronic t(9;22)-positive leukemias are more important than the initial cytogenetic aberration in determining chromosomal evolution. However, it is also entirely conceivable that submicroscopic differences between t(9;22) in acute and t(9;22) in chronic disease exist that influence the subsequent routes of evolution followed by the secondary changes.

Remarkable geographic heterogeneity has been noted in the distribution of the three characteristic secondary changes in CML (35). Among 581 cases with t(9;22) collected from six regions where at least 50 unselected and consecutively studied cases were available, the frequencies of $+8$, i(17q), and an additional Ph^1 were markedly different. Thus, the incidence of $+8$ varied from 21% in Sweden and New York to 76% in Japan; of i(17q) from 7% in Sweden to 28% in the U.S.S.R.; and of an extra Ph^1 from 7% in the U.S.S.R. to 54% in France.

The reason for these observed differences is completely unknown. It is tempting to speculate that variable exposure to exogenous factors may be of importance. Results from experimental carcinogenesis suggest that the etiologic factor may, under certain conditions, be expected to induce specific chromosomal patterns (40). Some observations indicate that such correlations may also exist in man: Differences in aberration pattern of malignant cells are found in patients who have been exposed to potential mutagenic/carcinogenic agents, compared to those without any known exposure (11,38), and in *de novo* ANLL, compared to ANLL secondary to treatment for another malignant disorder (11,52). It may in this context also be worthy of note that intensive chemotherapy during the chronic phase of CML has been reported to produce clones with preferential engagement of chromosome 1 (1). Therefore, differences in treatment among different countries might possibly be one reason for the observed heterogeneity. Obviously, the alternative to these explanations invoking environmental factors might be that differences in the genetic and ethnic backgrounds cause the variations. Unfortunately, the available material is insufficient to differentiate between these mechanisms.

VARIANT TRANSLOCATIONS IN Ph[1]-POSITIVE CML

The Ph[1]-marker chromosome originates through mechanisms other than t(9;22) in approximately 5% to 10% of CML cases (10,17,37,44,54). Traditionally, these variant translocations have been subdivided into two subgroups. In simple variant translocations, the deleted segment of $22q$ is translocated onto a chromosome other than chromosome 9, whereas three or more chromosomes are involved in complex variant translocations. An example of the complex variant is provided in Fig. 5a; a simple variant in Fig. 5b.

A total of 247 cases of CML with variant translocations are included in the tumor karyotype registry (34). The majority (151) were complex translocations; two were fiveways, nine were fourways, and the remaining 140 involved three chromosomes. Chromosome 9 was visibly involved in all but eight of the 151 cases. Simple variant translocations had been reported in 87 cases. Simultaneous involvement of chromosome 9 was apparent in only eight of these patients, taking part in translocations with seven different chromosomes. The distribution of breakpoints in both simple and complex variant translocations are provided in Fig. 6. The stan-

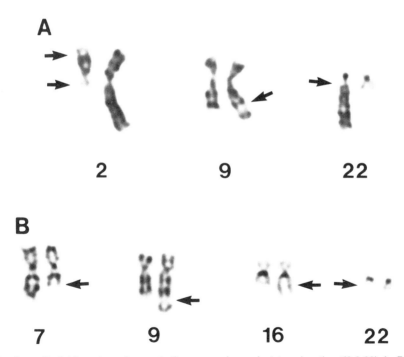

Fig. 5 a: Partial karyotype demonstrating a complex variant translocation t(2;9;22). **b**: Partial karyotype demonstrating a simple variant translocation t(16;22) with simultaneously occurring t(7;9).

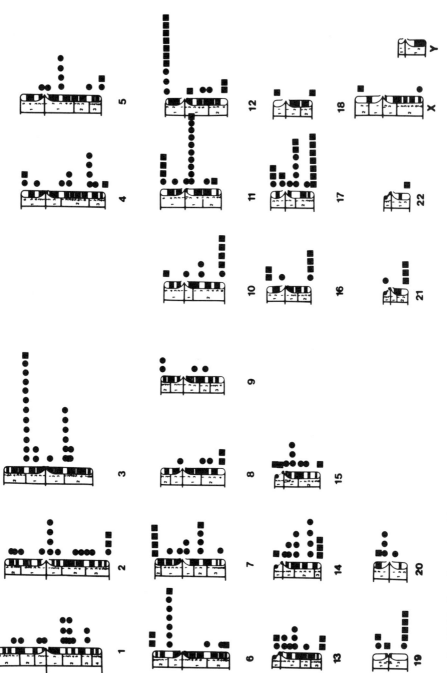

Fig. 6. Distribution of complex (●) and simple (■) variant translocation breakpoints in Ph¹-positive CML. The standard CML breakpoints 9q34 and 22q11 are not shown.

dard CML breakpoints 9q34 and 22q11 are not shown in Fig. 6. As can be seen, all chromosomes, with the single exception of the Y, have now been implicated in variant translocations in CML. However, the distribution pattern of the breakpoints is clearly nonrandom, with marked clustering to some chromosomal bands and regions (e.g., 3p21, 3q21, 11q13, 12p13, 17q25). The early impression of Sandberg (54) that variant breakpoints are primarily located in bands that stain lightly in G-banding is confirmed in the present, combined material: 85% of the translocation breakpoints in Fig. 6 affect light-staining bands, with only 15% in dark-staining regions. This may be compared with the distribution of breakpoints underlying structural aberrations in neoplasms with one single deviation from normality (33), where 66% of the affected bands were light-staining and 34% dark-staining.

Mitelman and Levan (37) have pointed out that simple variant translocation breakpoints much more often than breakpoints in complex translocations affect terminal bands. In Fig. 6, 90% of breakpoints in the simple rearrangements are terminal, whereas in complex translocations, 90% are interstitial.

Initially, no explanation for this peculiar difference in distribution was available. An investigation with high-resolution R-banding (this banding technique darkly stains bands that appear light-colored in G-banding and is consequently superior in detecting small structural rearrangements on the terminal tips of chromosomes) has demonstrated that apparently simple variant Ph^1-translocations in actual fact were complex rearrangements involving 9q34 (15). This conclusion, based on improved cytogenetic analysis, was confirmed by the same group when they demonstrated with the help of *in situ* hybridization that c-*abl* (a cellular oncogene normally located in 9q34) was moved to 22q, even in cases where initial examination using G- or Q-banding had failed to detect any rearrangement of 9q34 (6,15). Japanese investigators (24) have also reinvestigated cases previously classified as simple variant translocations using a combination of banding techniques and with special emphasis on possible minute, terminal rearrangements. In all the cases examined, the renewed scrutiny revealed that 9q34 was involved in the translocation, which should therefore be reclassified as complex. These data strongly indicate that simple variant translocations may be even less frequent findings than has been estimated in the past. Improved cytogenetic investigations utilizing banding techniques specifically suited to the detection of small terminal chromosome changes may reduce the number further and perhaps eventually demonstrate that all variant translocations in Ph^1-positive CML also involve chromosome 9. Seen against this background, the preference for terminal regions in simple variant translocations becomes perfectly understandable: The smaller the segments involved are, the easier it is to overlook the 9q34 involvement.

The additional changes in 41 Ph^1-positive CML patients with variant translocations are presented in Table 3, with a graphic presentation of the data in Fig. 7. Translocations involving chromosome 22 and/or 9 are not included. It is clear that the distribution of secondary changes is basically identical in classical and variant Ph^1-positive CML. Trisomy 8, i(17q), and $+Ph^1$ are the dominant features, with

TABLE 3. Secondary chromosome aberrations in 41 patients with Ph¹-positive CML and variant translocations[a]

Chromosome	+	−	t	del	inv	i	ins	der	dic	dup	r	Total	%
1	1		1	1								3	2.9
2												0	
3			2	1	1							4	3.9
4	1											1	0.9
5	1		2		2							5	4.9
6	2			1								3	2.9
7	1			1								2	1.9
8	13		3									16	15.5
9	2	2			1							5	4.9
10	2											2	1.9
11	3		1									4	3.9
12	1		1									2	1.9
13	1		1									2	1.9
14		1		1								2	1.9
15	1											1	0.9
16												0	
17	1		2			13						16	15.5
18	1	1	1									3	2.9
19	6	1										7	6.8
20			1									1	0.9
21	4		1									5	4.9
22		1		13				1				15	14.6
X		1										1	0.9
Y	1	2										3	2.9
mar												0	
dmin												0	
Total:	42	9	16	18	4	13		1				103	

[a]See Table 1 for explanation of abbreviations.

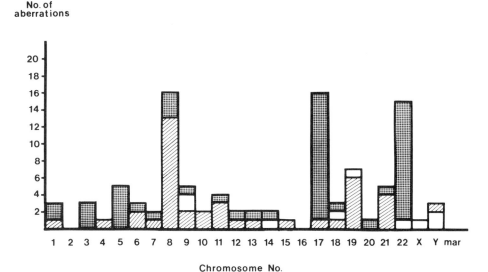

Fig. 7. Graphic distribution of secondary chromosome aberrations in Ph¹-positive CML with variant translocations: (*hatched area*) gain of whole chromosomes, (*open area*) loss of whole chromosomes; (*dotted area*) structural rearrangements.

trisomy 19 occurring somewhat less frequently. This great similarity in karyotypic evolution between these two CML groups indicates that the cytogenetic consequences of variant and classical Ph¹-producing translocations might be quite similar, an observation that is also supported by the fact that the two CML types are hematologically and prognostically indistinguishable (55).

CHROMOSOME ABERRATIONS IN Ph¹-NEGATIVE CML

We have reviewed all 95 Ph¹-negative CML patients included in the tumor karyotype registry (34). Chromosomal changes had been described in 65 of these patients, with an average of approximately two aberrations per case. The types of aberrations affecting the different chromosomes are provided in Table 4, with a graphic display of the information in Fig. 8. The prominent features trisomy 8 and i(17q), known from secondary aberrations in Ph¹-positive CML, are the most frequent cytogenetic hallmarks, even in Ph¹-negative CML. An extra Ph¹, of course, is not seen without a Philadelphia marker as the primary rearrangement. Some other differences between the Ph¹-negative CML and the secondary changes in CML associated with t(9;22) are also evident from a comparison of Figs. 3 and 8: Trisomy 19 is less frequent in Ph¹-negative leukemia. On the other hand, structural rearrangements of chromosome 3 and monosomy 7 seem to be common changes in this disease. The latter feature is reminiscent of the chromosome 7 involvement in acute t(9;22)-associated leukemias (18).

TABLE 4. Chromosome aberrations in 95 patients with Ph¹-negative CML[a]

Chromosome	+	−	t	del	inv	i	ins	der	dic	dup	r	Total	%
1			2	1						1		4	3.7
2			1		1							2	1.8
3			8	1								9	8.4
4			1									1	0.9
5				1								1	0.9
6			4									4	3.7
7		8	1									11	10.3
8	17		2	2								19	17.7
9	2	1	5									8	7.5
10												0	
11			3					1				4	3.7
12		1	1									1	0.9
13	3		1	1								6	5.6
14	1		1									2	1.8
15												0	
16												0	
17			1	2		8						11	10.3
18	1		1	1								3	2.8
19	2											2	1.8
20			1	3								4	3.7
21	5											5	4.7
22	3											3	2.8
X			1	1								1	0.9
Y		5	1									6	5.6
mar												0	
dmin												0	
Total:	34	15	34	13	1	8		1		1		107	

[a]See Table 1 for explanation of abbreviations.

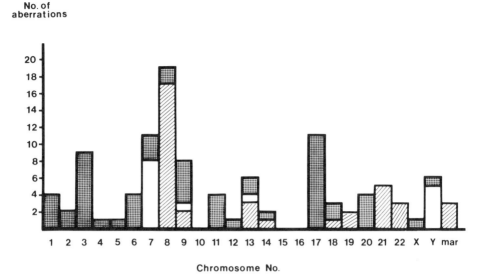

Fig. 8. Distribution of chromosome aberrations in Ph[1]-negative CML. (*hatched area*) gain of whole chromosomes; (*open area*) loss of whole chromosomes; (*dotted area*) structural rearrangements.

The relatively limited group of translocations engaging chromosome 9 in Ph[1]-negative CML merits more thorough consideration. All five rearrangements (Table 4) involve 9q34, the band consistently affected in the regular t(9;22) and now increasingly implicated even in Ph[1]-translocations, where previous investigations have failed to demonstrate any 9q34 involvement. It is possible that the growing awareness of the fundamental role of 9q34 rearrangement in leukemia pathogenesis (see below and elsewhere in this volume), combined with improved investigative techniques, may lead to the detection of minute aberrations affecting 9q34, even in Ph[1]-negative CML.

Lawler (29) has pointed out that whereas the average survival in Ph[1]-negative CML is short, there is a cohort of patients with this disease that have survival rates comparable to Ph[1]-positive patients. It could be speculated that this group of patients perhaps have 9q34 rearrangements pathogenetically equivalent to the Ph[1]-producing rearrangements in the classical and variant Ph[1]-positive CML.

CHROMOSOMAL ABERRATIONS IN CHRONIC MYELOID LEUKEMIA IN FUNCTIONAL PERSPECTIVE

In this presentation of the dynamic picture of cytogenetic aberrations in CML, we have consciously refrained from any in-depth discussion of the prognostic implications of the various abnormalities; however, a few important correlations may be briefly pointed out. Patients with classical and variant Ph^1-positive CML seem to have identical prognosis, and the course of the disease is also highly similar in these two cytogenetic CML subgroups (54). The average patient with Ph^1-negative CML, on the other hand, has a significantly shorter life expectancy (55). It should also be mentioned that during the quiescent chronic phase, when hematologic parameters are kept normal or near normal with only a minimum of cytostatic therapy, the Ph^1 marker still persists, illustrating that the seemingly benign picture at this stage by no means represents a true remission.

Current thinking about neoplasia visualizes the process of transformation as multistaged, with two or more mutational and/or epigenetic events acting as initiating and promoting influences in bringing about the final result: the malignant clone. The evidence behind this line of reasoning stems from both experimental and epidemiologic studies (25,46). The two- or multistep model of malignant transformation has been discussed in depth by Knudson (26,27). Although the theory was originally developed to account for the behavior of hereditary childhood malignancies, in particular autosomal dominant tumors, the two-hit hypothesis fits equally well the common sporadic cancers. Indeed, most human malignancies seem to have both a relatively common sporadic form and a rare, hereditary variant (28). CML is no exception in this context. In the vast majority of cases, the disease seems to strike completely at random, but a few conspicuous familial aggregates of Ph^1-positive CML have been described (21,59,61). These reports are compatible with an uncommon form of the disease segregating in an autosomal dominant manner.

Chronic myeloid leukemia, displaying consistent chromosomal changes throughout the course of the disease, would from a cytogenetic point of view seem to fit a two- or multihit hypothesis of malignant transformation admirably. The highly regular Ph^1 abnormality occurs very early in the disease and could be viewed as the cytogenetic manifestation of the initiating event. The crucial question is: Is t(9;22), or a functionally equivalent variant translocation, the abnormality that produces the malignancy or is the translocation secondary to an earlier submicroscopic change that is sufficiently leukemogenic in its own right?

No satisfactory evidence is available to answer conclusively this question either way. Until recently, the evidence for a pathogenetically important role of t(9;22) in CML was purely cytogenetic: The fact that this abnormality is such a specific and regular feature of CML argues forcefully that it is an important step in leukemogenesis. Likewise, the finding of Ph^1-positive bone marrow cells preceding the leukemia (3) adds additional support to this interpretation.

The past few years of research employing molecular genetic methods have opened up new avenues of understanding with regard to the molecular consequences of chromosome aberrations in cancer. The biochemical evidence for a pathogenetic role of t(9;22) in CML is reviewed extensively by Groffen elsewhere in this volume. It suffices here to summarize but a few especially pertinent items of information from this field: The cellular oncogene c-*abl* is translocated from 9q34 to 22q, not only in classical CML (6), but also in variant translocations where chromosome 9 seems unaffected under microscopic examination (15). An abnormal *abl-bcr* transcript has now been demonstrated in CML cells (19,57). Although the function of this abnormal product is still largely unknown, it seems a very attractive possibility that it plays at least some active part in the neoplastic process. Hence, evidence for the direct, causal importance of translocations involving 9q34 in CML is now based not solely on cytogenetic findings, but is independently supported by molecular genetic investigations on oncogene rearrangements.

It must be emphasized, however, that in spite of the attractiveness of the translocation–oncogene rearrangement hypothesis outlined above, the evidence favoring this model is still only circumstantial. Furthermore, some observations seem to be at odds with the conclusion that t(9;22) is the salient leukemogenic event in CML.

Fialkow et al. (9) have demonstrated that the Ph^1 in CML is present not only in granulocytes and their progenitors, but even in other bone marrow cell lineages. Working with the G-6PD marker, they detected that erythrocytes, platelets, eosinonophils, monocytes, B-lymphocytes, and granulocytes were part of the same clone and drew the conclusion that the basic change in CML probably involves a stem cell multipotent for all these elements. The same group has also utilized "long term" human bone marrow cultures to find that some blood lymphocytes arising from Ph^1-positive CML clones were Ph^1-negative (8). This led to the hypothesis that at least two steps were involved in leukemogenesis; one causing growth of a clone of pluripotent hematopoietic cells, the other inducing Ph^1 in descendants of these progenitors (7,8). Compatible with this suggestion that Ph^1 formation may not be an important, let alone the sole important, pathogenetic event in CML are the reports of six patients altogether with CML who developed Ph^1 only late in their disease (16,31,32,58). To incorporate these data with the main bulk of cytogenetic evidence outlined previously, a three-step scheme for the pathogenesis of CML has been suggested (7):

Step 1. Clonal proliferation of pluripotent Ph^1-negative bone marrow stem cells. These are genetically unstable and have a growth advantage over normal stem cells. This step, which involves a still totally unknown primary event, might be equivalent to initiation in experimental tumorigenesis.

Step 2. Acquisition of Ph^1 (or a functionally equivalent cytogenetic rearrangement). Perhaps this event has to take place in a suitably primed cell, presumably a pluripotent stem cell, for CML to evolve. The key effect in molecular terms of the translocation may be a deranged c-*abl* function. This stage might be analogous to tumor promotion in experimental systems.

Step 3. Blast crisis. The increased genetic instability now results in additional chromosomal abnormalities, thereby enhancing the malignant potential of the abnormal clone. This stage might cover the same development which in experimental carcinogenesis is termed tumor progression.

Regardless of whether the primary event in CML is a cytogenetically recognizable rearrangement or a submicroscopic leukemia-inducing change preceding the characteristic chromosome aberration, we still do not know what causes the event. If we address only the change observable by cytogenetic methods, why is it that t(9;22) is the rearrangement occurring with such regularity in this disease, even when a number of other changes quite obviously are compatible with the same biological consequences on both molecular and cellular levels? A selection hypothesis assuming that different translocations occur at random and with comparable frequencies hardly seems sufficient to account for the observed data. Instead, the specific, acquired chromosome abnormalities in cancer, of which t(9;22) is a prime example, may perhaps be actively induced, the result of a direct interaction between a mutagen and the genomic sites liable to participation in the subsequent rearrangement (36). Rowley (50) has pointed at two factors that might favor such active interaction: proximity between the two chromosome regions within the interphase nucleus, and regions of homology in these two neighboring chromosomes. However, techniques are not yet available to test the relevance and accuracy of these hypotheses.

The rapidly expanding amount of data with regard to the functioning of oncogenes in normal and malignant cells and the very consistent cytogenetic picture in CML enable a combination of these two fields of knowledge to provide a reasonable tentative picture of the biological consequences of the typical chronic phase chromosome aberration. Similar attempts at correlations are more difficult at the very outbreak of disease or with regard to immediately premorbid bone marrow cells, where the pathogenetic importance of t(9;22) is still completely unresolved. The realization in the very recent past that the genetic material in 9q34 may be more important than genes in 22q in this disease (19,57), and especially the fact that renewed examinations have found 9q34 to be rearranged even in such cases where the chromosome in the initial study seemed cytogenetically normal (15), cautions that the few patients reported in the past (16,31,32,58) with late-appearing Ph^1 markers may still have had subtle rearrangements affecting 9q34. In the future, *in situ* hybridization using radioactive probes against c-*abl* might help to clarify the evolutionary process in such patients.

At the other end of the disease spectrum, in blast-phase CML, the complexity of the secondary changes makes the functional implications of the various aberrations hard to unravel. The data indicate that the initial change, i.e., t(9;22), is a most important factor in determining the subsequent cytogenetic evolution of the malignant clone (Table 1, Fig. 3). This element of cytogenetic predetermination of secondary aberrations by the primary abnormality is certainly not incompatible with the notion of selective pressures playing important parts in ultimately deciding which subclones will eventually prevail in the evolutionary competition among variously deranged cells. The mechanisms whereby secondary changes increase the malignant potential of the clone are not known. An amplification of the initial, primary event might be one way of achieving this (36) and would of course explain the propensity for acquiring an extra Ph^1 marker in the later disease stages. The reasons for an extra chromosome 8 or 19 or i(17q) or certain combinations thereof conferring on the subclone an evolutionary edge are at present almost totally inscrutible. The presence on 17q of the oncogenes c-*erbA* and c-*neu* and on 8q of c-*myc* and c-*mos* (22,56) may well be of importance in this context. Some observations from comparative oncogenetics are quite intriguing in relation to the question of oncogenes and cancer-associated trisomies (25): Trisomy 7 is the most common chromosomal change in Rous sarcoma virus-induced rat tumors, and c-*myc* is localized on chromosome 7 in the rat. Likewise, trisomy 15 is the most common and often single chromosomal anomaly in mouse leukemia, and c-*myc* is located on chromosome 15 in the mouse. Hence, the parallel between these two animal systems and trisomy for the *myc*-containing chromosome 8 in man is very suggestive.

Investigative efforts combining cytogenetic and molecular genetic techniques may in the not too distant future substantially increase our understanding of the dynamic interplay between genomic changes and their phenotypic molecular effects in both the initial and later stages of CML. A more profound appreciation of the crucial pathogenetic events involved will provide the best basis for therapeutic approaches more directly aimed at the underlying abnormality than are presently available.

ACKNOWLEDGMENT

Original research reported in this chapter was supported by the Swedish Cancer Society and the J.A.P. Foundation for Medical Research.

REFERENCES

1. Alimena, G., Brandt, L., Dallapiccola, B., Mitelman, F., and Nilsson, P.G. (1979): Secondary chromosome changes in chronic myeloid leukemia: Relation to treatment. *Cancer Genet. Cytogenet.*, 1:79–85.
2. Alimena, G., Dallapiccola, B., Gastaldi, R., Mandelli, F., Brandt, L., Mitelman, F., and Nilsson, P.G. (1982): Chromosomal, morphological and clinical correlations in blastic crisis of chronic myeloid leukaemia. A study of 69 cases. *Scand. J. Haematol.*, 28:103–117.
3. Canellos, G.P., and Whang-Peng, J. (1972): Philadelphia chromosome-positive preleukaemic state. *Lancet*, 2:1227–1228.
4. Caspersson, T., Gahrton, G., Lindsten, J., and Zech, L. (1970): Identification of the Philadelphia chromosome as a number 22 by quinacrine mustard fluorescence analysis. *Exp. Cell Res.*, 63:238–240.
5. de Grouchy, J., de Nava, C., Cantu, J.M., Bilski-Pasquier, G., and Bousser, J. (1966): Models for clonal evolutions: a study of chronic myelogenous leukemia. *Am. J. Hum. Genet.*, 18:485–503.
6. de Klein, A., van Kessel, A., Grosveld, G., Bartram, C.R., Hagemeijer, A., Bootsma, D., Spurr, N.K., Heisterkamp, N., Groffen, J., and Stephenson, J.R. (1982): A cellular oncogene is translocated to the Philadelphia chromosome in chronic myelogenous leukemia. *Nature*, 300:765–767.
7. Fialkow, P.J. (1984): Clonal evolution of human myeloid leukemias. In: *Genes and Cancer*, edited by J.M. Bishop, J.D. Rowley, and M. Greaves, pp. 215–226. Alan R. Liss, New York.
8. Fialkow, P.J., Denman, A.M., Jacobson, R.J., and Lowenthal, M.N. (1978): Chronic myelocytic leukemia: origin of some lymphocytes from leukemic stem cells. *J. Clin. Invest.*, 62:815–823.
9. Fialkow, P.J., Jacobson, R.J., and Papayannopoulou, T. (1977): Chronic myelocytic leukemia:

Clonal origin in a stem cell common to the granulocyte, erythrocyte, platelet and monocyte/macrophage. *Am. J. Med.*, 63:125–130.
10. First International Workshop on Chromosomes in Leukemia. (1978): Chromosomes in Ph[1]-positive chronic granulocytic leukaemia. *Br. J. Haematol.*, 39:305–309.
11. Fourth International Workshop on Chromosomes in Leukemia, 1982 (1984): A prospective study of acute non-lymphocytic leukemia. *Cancer Genet. Cytogenet.*, 11:249–360.
12. Gahrton, G., Lindsten, J., and Zech, L. (1974): The Philadelphia chromosome and chronic myelocytic leukemia (CML)—Still a complex relationship? *Acta Med. Scand.*, 196:353–354.
13. Groffen, J., Stephenson, J.R., Heisterkamp, N., de Klein, A., Bartram, C.R., and Grosveld, G. (1984): Philadelphia chromosomal breakpoints are clustered within a limited region, bcr, on chromosome 22. *Cell*, 36:93–99.
14. Gödde-Salz, E., Schmitz, N., and Bruhn, H.-D. (1985): Philadelphia chromosome (Ph) positive chronic myelocytic leukemia (CML): Frequency of additional findings. *Cancer Genet. Cytogenet.*, 14:313–323.
15. Hagemeijer, A., Bartram, C.R., Smit, E.M.E., van Agthoven, A.J., and Bootsma, D. (1984): Is the chromosomal region 9q34 always involved in variants of the Ph[1] translocation? *Cancer Genet. Cytogenet.*, 13:1–16.
16. Hayata, I., Sakurai, M., Kakati, S., and Sandberg, A.A. (1975): Chromosomes and causation of human cancer and leukemia. XVI. Banding studies of chronic myelocytic leukemia, including five unusual Ph[1] translocations. *Cancer*, 36:1177–1191.
17. Heim, S., Billström, R., Kristoffersson, U., Mandahl, N., Strömbeck, B., and Mitelman, F. (1985): Variant Ph translocations in chronic myeloid leukemia. *Cancer Genet. Cytogenet.*, 18:215–227.
18. Heim, S., and Mitelman, F. (1986): Secondary chromosome aberrations in the acute leukemias. *Cancer Genet. Cytogenet.*, 22:331–338.
19. Heisterkamp, N., Stam, K., and Groffen, J. (1985): Structural organization of the bcr gene and its role in the Ph[1] translocation. *Nature*, 315:758–761.
20. Heisterkamp, N., Stephenson, J.R., Groffen, J., Hansen, P.F., de Klein, A., Bartram, C.R., and Grosveld, G. (1983): Localization of the c-abl oncogene adjacent to a translocation breakpoint in chronic myelocytic leukemia. *Nature*, 306:239–242.
21. Hirschhorn, K. (1968): Cytogenetic alterations in leukemia. In: *Perspectives in Leukemia*, edited by W. Dameshek and R.M. Dutcher, pp. 113–120. Grune & Stratton, New York.
22. Human Gene Mapping 8. Helsinki Conference (1985): *The 8th International Workshop on Human Gene Mapping*. S. Karger AG, Basel.
23. ISCN (1978): An International System for Human Cytogenetic Nomenclature (1978). *Birth Defects: Original Article Series*, Vol. XIV, No. 8. The National Foundation, New York.
24. Ishihara, T., Minamihisamatsu, M., and Tosuji, H. (1985): Chromosome 9 in variant Ph translocations. *Cancer Genet. Cytogenet.*, 14:183–184.
25. Klein, G., and Klein, E. (1985): Evolution of tumours and the impact of molecular oncology. *Nature*, 315:190–195.
26. Knudson, A.G. (1971): Mutation and cancer: Statistical study of retinoblastoma. *Proc. Natl. Acad. Sci. U.S.A.*, 68:820–823.
27. Knudson, A.G. (1977): Genetic and environmental interaction in the origin of human cancer. In: *Genetics of Human Cancer*, edited by J.J. Mulvihill, R.W. Miller, and J.F. Fraumeni, Jr. pp. 391–399. Raven Press, New York.
28. Knudson, A.G., Strong, L.C., and Anderson, D.E. (1973): Heredity and cancer in man. *Prog. Med. Genet.*, 9:113–158.
29. Lawler, S.D. (1982): Significance of chromosome abnormalities in leukemia. *Semin. Hematol.*, 19:257–272.
30. Lawler, S., O'Malley, F., and Lobb, D.S. (1976): Chromosome banding studies in Philadelphia chromosome positive myeloid leukaemia. *Scand. J. Haematol.*, 17:17–28.
31. Lisker, R., Casas, L., Mutchinick, O., Lopez-Ariza, B., and Labardini, J. (1982): Patient with chronic myelogenous leukemia and late appearing Philadelphia chromosome. *Cancer Genet. Cytogenet.*, 6:275–277.
32. Lisker, R., Casas, L., Mutchinick, O., Perez-Chavaz, F., and Labardini, J. (1980): Late appearing Philadelphia chromosome in two patients with chronic myelogenous leukemia. *Blood*, 56:812.
33. Mitelman, F. (1984): Restricted number of chromosomal regions implicated in aetiology of human cancer and leukaemia. *Nature*, 310:325–327.

34. Mitelman, F. (1985): *Catalog of Chromosome Aberrations in Cancer*, 2nd ed. Alan R. Liss, Inc., New York.
35. Mitelman, F. (1986): Geographic heterogeneity of chromosome aberrations in hematologic disorders. *Cancer Genet. Cytogenet.*, 20:203–208.
36. Mitelman, F., and Levan, G. (1978): Clustering of aberrations to specific chromosomes in human neoplasms. III. Incidence and geographic distribution of chromosome aberrations in 856 cases. *Hereditas*, 89:207–232.
37. Mitelman, F., and Levan, G. (1981): Clustering of aberrations to specific chromosomes in human neoplasms. IV. A survey of 1,871 cases. *Hereditas*, 95:79–139.
38. Mitelman, F., Brandt, L., and Nilsson, P.G. (1978): Relation among occupational exposure to potential mutagenic/carcinogenic agents, clinical findings, and bone marrow chromosomes in acute nonlymphocytic leukemia. *Blood*, 52:1229–1237.
39. Mitelman, F., Levan, G., Nilsson, P.G., and Brandt, L. (1976): Non-random karyotypic evolution in chronic myeloid leukemia. *Int. J. Cancer*, 18:24–30.
40. Mitelman, F., Mark, J., Levan, G., and Levan, A. (1972): Tumor etiology and chromosome pattern. *Science*, 176:1340–1341.
41. Mitelman, F., Nilsson, P.G., and Brandt, L. (1975): Abnormal clones resembling those seen in blast crisis arising in the spleen in chronic myelocytic leukemia. *J. Natl. Cancer Inst.*, 54:1319–1321.
42. Nowell, P.C., and Hungerford, D.A. (1960): A minute chromosome in human granulocytic leukemia. *Science*, 132:1497.
43. O'Riordan, M.L., Robinson, J.A., Buckton, K.E., and Evans, H.J. (1971): Distinguishing between the chromosomes involved in Down's syndrome (trisomy-21) and chronic myeloid leukaemia (Ph[1]) by fluorescence. *Nature*, 230:167–168.
44. Oshimura, M., Ohyashiki, K., Terada, H., Takaku, F., and Tonomura, A. (1982): Variant Ph[1] translocations in CML and their incidence, including two cases with sequential lymphoid and myeloid crises. *Cancer Genet. Cytogenet.*, 5:187–201.
45. Pedersen, B. (1969): *Cytogenetic Evolution in Chronic Myelogenous Leukaemia. Relation of Chromosomes to Progression and Treatment of the Disease*. (Faculty of Medicine Dissertation). Munksgaard, Copenhagen.
46. Peto, R. (1982): Carcinogenesis as a multistage process-evidence from human studies. In: *Host Factors in Human Carcinogenesis* (IARC Sci. Publ. No. 39), edited by H. Bartsch and B. Armstrong. pp. 27–28. International Agency for Research on Cancer, Lyon.
47. Prakash, O., and Yunis, J.J. (1984): High resolution chromosomes of the t(9;22) positive leukemias. *Cancer Genet. Cytogenet.*, 11:361–367.
48. Prieto, F., Egozcue, J., Forteza, G., and Marco, F. (1970): Identification of the Philadelphia (Ph[1]) chromosome. *Blood*, 35:23–38.
49. Rowley, J.D. (1973): A new consistent chromosomal abnormality in chronic myelogenous leukaemia identified by quinacrine fluorescence and Giemsa staining. *Nature*, 243:290–293.
50. Rowley, J.D. (1977): A possible role for nonrandom chromosomal changes in human hematologic malignancies. In: *Chromosomes Today*, edited by A. de la Chapelle and M. Sorsa, Vol. 6 pp. 345–355. Elsevier/North Holland Biomedical Press, Amsterdam.
51. Rowley, J.D. (1980): Ph[1]-positive leukaemia, including chronic myelogenous leukaemia. *Clin. Haematol.*, 9:55–86.
52. Rowley, J.D., Golomb, H., and Vardiman, J.W. (1981). Nonrandom chromosome abnormalities in acute leukemia and dysmyelopoietic syndromes in patients with previously treated malignant disease. *Blood*, 58:759–767.
53. Sadamori, N., Matsunaga, M., Yao, E.-I., Ichimaru, M., and Sandberg, A.A. (1985): Chromosomal characteristics of chronic and blastic phases of Ph-positive chronic myeloid leukemia. *Cancer Genet. Cytogenet.*, 15:17–24.
54. Sandberg, A.A. (1980): Chromosomes and causation of human cancer and leukemia: XL. The Ph[1] and other translocations in CML. *Cancer*, 46:2221–2226.
55. Sandberg, A.A. (1980): *The Chromosomes in Human Cancer and Leukemia*. Elsevier/North Holland, New York.
56. Schechter, A.L., Hung, M-C., Vaidyanathan, L., Weinberg, R.A., Yang-Feng, T.L., Francke, U., Ullrich, A., and Coussens, L. (1985): The *neu* gene: an *erbB*-homologous gene distinct from and unlinked to the gene encoding the EGF receptor. *Science*, 229:976–978.
57. Shtivelman, E., Lifshitz, B., Gale, R.P., and Canaani, E. (1985): Fused transcript of *abl* and *bcr* genes in chronic myelogenous leukaemia. *Nature*, 315:550–554.

58. Tanzer, J. (1977): Les anomalies chromosomiques dans les syndromes myeloproliferatifs chroniques. In: *Actualités Hematologiques*, Ser. 11, pp. 12–28. Masson, Paris.
59. Tokuhata, G.K., Neely, C.L., and Williams, D.L. (1968): Chronic myelocytic leukemia in identical twins and a sibling. *Blood*, 31:216–225.
60. Third International Workshop on Chromosomes in Leukemia (1980): Chromosomal abnormalities in acute lymphoblastic leukemia: Structural and numerical changes in 234 cases. *Cancer Genet. Cytogenet.*, 4:101–110.
61. Weiner, L. (1965): A family with high incidence leukemia and unique Ph^1 chromosome findings. *Blood*, 26:871.
62. Yunis, J.J. (1981): New chromosome techniques in the study of human neoplasia. *Hum. Pathol.*, 12:540–549.

Activation of c-*abl* as a Result of the Ph' Translocation in Chronic Myelocytic Leukemia

J. Groffen, N. Heisterkamp, and K. Stam

Laboratory of Molecular Genetics, Oncogene Science Inc., Mineola, New York

Evidence for the direct involvement of a family genes, called oncogenes, in human tumorigenesis is rapidly accumulating. Of the approximately 20 oncogenes identified to date, several have been associated with specific forms of human cancer. Oncogenes are believed to cause tumorigenesis either individually or by acting in concert; the interaction of oncogene-encoded proteins possibly accounts for the multistage nature of cancer. Because the rate of discovery of new oncogenes has markedly decreased during the past few years, it is currently expected that the total number of oncogenes will not expand much beyond 25 to 30. Some oncogenes, such as the individual members of the *ras* oncogene family, appear to be involved in the large majority of human cancers. In contrast, the subject of this chapter, c-*abl*, has been strongly associated with one specific type of human cancer, chronic myelocytic leukemia (CML).

Oncogenes are present in the human genome as normal cellular genes (proto-oncogenes) with normal cellular functions. Proto-oncogene amino acid sequences have been conserved throughout vertebrate evolution, suggesting that they perform a critical role in normal cellular metabolism. Many of the oncogenes for which the normal cellular function has been unraveled to date appear to represent genes that encode either growth factors or growth factor membrane receptors.

In general, oncogene activation embodies either qualitative or quantitative changes in the protein product encoded by an altered proto-oncogene. In addition, the activation could be caused by the expression of a normal oncogene product in normal quantities in the wrong tissues or at an inappropriate moment. The cellular processes modified by the conversion of a proto-oncogene to a transforming activated oncogene are not fully understood. A number of different genetic mechanisms have been shown to result in oncogene activation. These include point mutations, the substitution of a single nucleotide resulting in an altered amino acid in the oncogene-encoded protein; translocations, the shifting of DNA segments from one chromosome to another; gene rearrangement, insertion or deletion of DNA fragments within a gene; and gene amplification, increase in oncogene DNA copy number.

One of the most frequently identified chromosomal abnormalities associated with human tumorigenesis is the Philadelphia chromosome (Ph'). This chromosomal abnormality can be found in the leukemic cells of at least 95% of all patients with CML (40). The Ph' chromosome is the result of a reciprocal translocation in which parts of chromosomes 9 and 22 are exchanged (Fig. 1). The disease is characterized by an accumulation of myeloid cells and their precursors; in a more advanced stage, Ph'-positive CML cells are blocked in differentiation; the accumulation of undifferentiated blast cells will directly or indirectly be responsible for the almost invariably fatal consequences of this type of leukemia. In 1982, the human c-*abl* proto-oncogene was cloned and localized on chromosome 9 (22,23). The latter finding implied that c-*abl* could have moved from chromosome 9 to 22 in the Ph' translocation. Indeed, subsequent experiments demonstrated such relocation of human c-*abl* (2,11), even in CML patients with complex Ph' translocations.

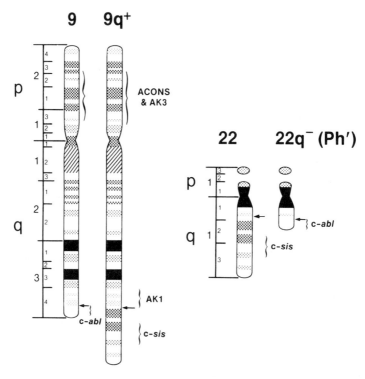

Fig. 1. Diagramatic representation of the involvement of c-*abl* in the Ph' translocation. Map positions of ACONS, AK3, and AK1, and c-*sis* (1) are indicated. Localization of c-*abl* within the terminal portion of chromosome 9 (*q*34) is as described in the text. The *arrows* point to the breakpoints on chromosome 9 (c-*abl*) and on chromosome 22 (*bcr*).

Since then, data concerning the molecular basis of CML have accumulated rapidly: On chromosome 22, a "hot-spot" region containing the translocation breakpoints of the Ph' chromosome could be determined. We have proposed the name "breakpoint cluster region" or *bcr* for this area on chromosome 22 (21). In addition, abnormally sized c-*abl* mRNA and protein products were found in CML cell lines and in the leukemic cells of patients. These products were thought to be chimeric *bcr-abl* gene products, and final proof for the presence of a chimeric *bcr*/c-*abl* mRNA has indeed been provided (24,44). The association between the presence of this abnormal mRNA and the occurrence of one type of human leukemia, CML, is one of the strongest to date, providing a link between the activation of an oncogene and tumorigenesis.

HUMAN c-*abl*

Abelson murine leukemia virus (AbMuLV) is a recombinant between Moloney murine leukemia virus and cellular sequences of mouse origin designated v-*abl* (17). The human genome contains sequences highly homologous to the mouse v-*abl* and thus exhibits a strong degree of conservation in evolution—the human c-*abl* oncogene. Because the v-*abl* homologous sequences within the human oncogene span a region of at minimum 30 kilobases (kb), they were cloned from a human cosmid library (22).

As found in AbMuLV, v-*abl* sequences are without introns and may be regarded as incomplete cDNA copies of the murine c-*abl* oncogene: Whereas v-*abl* is approximately 3 kb in length (37,38,17), v-*abl* homologous mRNAs of 5 to 7 kb (mouse) (46) and 6 to 7 kb (human) have been detected. In human, v-*abl* homologous sequences are distributed discontinuously over a region of 32 kb and are dispersed over at least 9 exons (22). On isolation of a human c-*abl* cDNA of 6 kb (44), and submitted for publication two additional exons were identified 5' to the v-*abl* homologous exons: One is located 5' to the most 5' v-*abl* homologous exon (A in Fig. 2A); the second is located 5' of the most 5' Bam HI site (Fig. 2A). Although an ATG codon is found within this latter exon, it has not been established conclusively whether the c-*abl* oncogene utilizes this start codon or contains additional coding sequences to the 5' (44). The gene spans a region of cellular DNA of at minimum 48 kb.

Southern hybridization of human c-*abl* probes to DNAs of mouse/human somatic cell hybrids that have retained specific human chromosomes established the chromosomal location of c-*abl* at human chromosome 9 (23). These findings were of interest because a highly specific translocation, the Philadelphia translocation (Ph'), has been found in the leukemic cells over 95% of all CML patients, involving an exchange of parts of the long arms of chromosomes 9 and 22 (40) (Fig. 1). The location of c-*abl* on chromosome 9 implies that this oncogene could be involved in CML; therefore, we investigated the chromosomal location of the c-*abl* gene in the leukemic cells of Ph'-positive CML patients. Southern blot analyses of the DNAs of somatic cell hybrids in which the $9q^+$ and $22q^-$ (or Ph') chromosome had segregated, revealed that c-*abl* had been translocated to the Ph' chromosome (11). This result was confirmed by *in situ* hybridization; it established, that the Philadelphia translocation involves a reciprocal exchange of genetic material between chromosomes 9 and 22. The segment of chromosome 9 translocated to chromosome 22 is too small to be detected by cytogenetic analysis; it encompasses chromosome 9q34 to 9qter and allows the localization of c-*abl* to the most telomeric region of the long arm of chromosome 9.

Although the majority of Ph'-positive CML patients have a t(9;22) (q34;q11) translocation, 6 to 8% have variant cytogenetic forms. Two such patients with complex translocations, a t(9;11;22) and a t(1;9;22) were examined by in *situ* hybridization and Southern blot analysis of DNA of somatic cell hybrids for the localization of the c-*abl* oncogene. In both patients, c-*abl* had been translocated to chromosome 22 and was thus situated on the Ph' chromosome (1).

Breakpoint on Chromosome 9

These results indicated strongly, the the c-*abl* oncogene could be involved in Ph'-positive CML. If c-*abl* were actively involved in the development of this disease, one would expect that the breakpoint on chromosome 9 would occur either within or in relatively close proximity to human c-*abl*. To investigate this possibility, high-molecular-weight DNA was isolated from biopsy samples of three CML patients. Each DNA was digested with Bam HI and subjected to Southern blot analysis. Extensive analysis with probes isolated from the v-*abl* homologous region of c-*abl* revealed no abnormalities or rearrangements in the DNAs of these patients. A probe for the Bam HI fragment located immediately 5' to the v-*abl* homologous exons (a 5' 0.2-kb Bam HI/EcoRI fragment, Fig. 2) detected a normal 14.5-kb Bam HI fragment in normal human DNA and in the DNAs of the three CML patients; however, in the DNA of one of the patients, patient 0319129, an additional Bam HI restriction fragment approximately 3.1 kb was identified (data not shown). The most likely explanation for this finding is that one allelic copy of DNA sequences immediately 5' of the v-*abl* homologous sequences in patient 0319129 is normal, whereas the second is rearranged. Molecular cloning of the rearranged sequences confirmed this hypothesis (Fig. 2B). Sequences in the 5' region of this fragment (the black bar) completely match with the c-*abl* locus on chromosome 9. In contrast, sequences more to the 3' have no homology to chromosome 9 sequences. To determine the origin of the latter sequences, a 1.2-kb Hind III/Bgl II probe (1.2 HBg) was prepared (Fig. 2B). Under stringent washing conditions, this probe hybridized to a 5.0-kb Bgl II fragment in normal human DNA, but not to sequences in mouse DNA or in rodent/human somatic cell hybrids containing human chromosome 9. However, this probe detected a 5.0-kb Bgl II fragment in a somatic cell hybrid containing chromosome 22 sequences as its only human component. Thus, the cloned rearranged fragment of patient 0319129

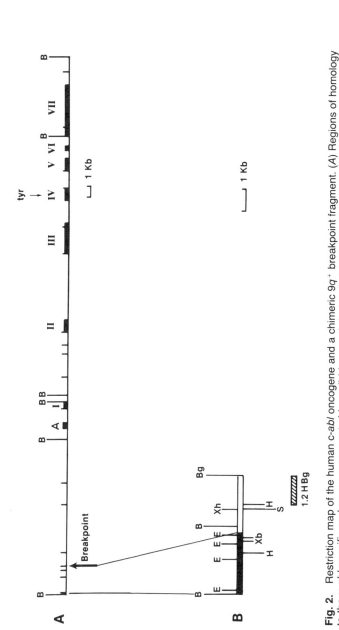

Fig. 2. Restriction map of the human c-*abl* oncogene and a chimeric 9q+ breakpoint fragment. (A) Regions of homology to the v-*abl* specific probes are represented by *solid bars* and designated I–VII. The first 5′ non-v-*abl* homologous c-*abl* exon is indicated by A. EcoR I sites are marked by *small vertical lines*. The *vertical arrow* points to the breakpoint in the DNA of CML 0319129. The chimeric fragment in B is a subclone of a 6.0-kb Bg III fragment isolated from CML patient 0319129. The *solid bar* indicates sequences from chromosome 9; the *open bar* represents chromosome 22 sequences. Restriction enzymes include Bam HI (B); Bgl II (Bg); Bst EII (Bs); Hind III (H); Sst I (S); Xba I (Xb); Xho I (Xh); Eco RI (E); and Kpn I (K). The phosphotyrosine acceptor region is indicated with *tyr*.

represents a chimeric chromosome $9q^+$ breakpoint fragment containing sequences originating from chromosomes 9 and 22 (25).

Within the c-*abl* oncogene, the breakpoint falls within a large intron of 16 kb, separating an exon 5' of the 5' Bam HI site from the exon designated as A (Fig. 2A). Although two other breakpoints on chromosome 9 have been reported to be situated within this large intron (31), breakpoints on chromosome 9 are spread over a relatively extended region of DNA of over 90 kb, 5' of the most 5'-identified c-*abl* exon; thus, although three breakpoints on chromosome 9 are located within c-*abl*, other breakpoints may be situated 5' to the gene itself. Nonetheless, the localization of some breakpoints within c-*abl* and of others in relative proximity provides evidence for a role of c-*abl* in the generation of CML.

IDENTIFICATION OF A BREAKPOINT CLUSTER REGION ON CHROMOSOME 22

The molecular cloning of a chimeric restriction enzyme fragment containing the t(9;22) breakpoint from patient 0319129 enabled us to use the chromosome 22 sequences as a molecular probe. Hybridization of the chromosome 22-specific 1.2-kb HBg probe to DNA isolated from the leukemic spleen of a second CML patient, 02120185, revealed the presence of abnormal restriction fragments. This finding suggested that in both CML patients a breakpoint had occurred within a limited stretch of chromosome 22 DNA that could be examined with this probe. To investigate this possibility, we cloned the chromosome 22 sequences corresponding to the 1.2-kb HBg probe from a normal human cosmid library (Fig. 3A). Using the 1.2-kb HBg prove and a 0.6-kb Hind III/Bam HI probe situated 5' to it (Fig. 3A), we have investigated the presence of breakpoints in more than 40 DNAs isolated from the leukemic cells (including blood, bone marrow, and spleen) of Ph'-positive CML patients. All patients examined to date were found to have a breakpoint within a 5.8-kb region on chromosome 22, for which we have proposed the name *breakpoint cluster region* or *bcr* (21).

In CML, more than 95% of patients have a Philadelphia chromosome; the remaining Ph'-negative patients generally have a poorer prognosis. Since no Ph' chromosome can be detected cytogenetically, one would not expect to find the translocation of c-*abl* to chromosome 22 or a break occurring within *bcr*. Indeed, in two cases of Ph'-negative CML with apparently normal karyotypes, c-*abl* remained on chromosome 9 (2). In two different patients, no rearrangements were visible within *bcr* on chromosome 22 (21). Although the following conclusion therefore seems justified, that Ph'-positive patients contain a translocated c-*abl* oncogene and a breakpoint in *bcr*, whereas Ph'-negative patients do not, the lack of a cytogenetically visible Ph' chromosome is not an unambiguous indication for the lack of involvement of *bcr* and c-*abl*: In a Ph'-negative CML patient with a t(9;12) ($q34;q21$) and two apparently normal chromosomes 22, *bcr* and c-*abl* were found to be jointly translocated to chromosome 12; this patient had an unusually long clinical remission compared to other Ph'-negative patients (3).

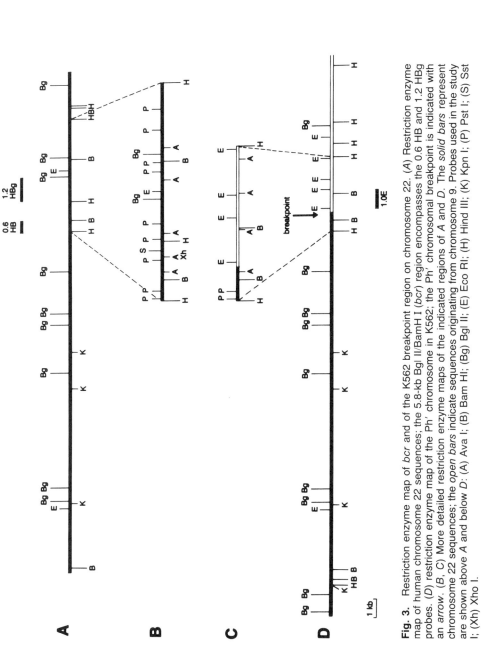

Fig. 3. Restriction enzyme map of *bcr* and of the K562 breakpoint region on chromosome 22. (A) Restriction enzyme map of human chromosome 22 sequences; the 5.8-kb Bgl II/BamH I (*bcr*) region encompasses the 0.6 HB and 1.2 HBg probes. (D) restriction enzyme map of the Ph' chromosome in K562; the Ph' chromosomal breakpoint is indicated with an *arrow*. (B, C) More detailed restriction enzyme maps of the indicated regions of A and D. The *solid bars* represent chromosome 22 sequences; the *open bars* indicate sequences originating from chromosome 9. Probes used in the study are shown above A and below D: (A) Ava I; (B) Bam HI; (Bg) Bgl II; (E) Eco RI; (H) Hind III; (K) Kpn I; (P) Pst I; (S) Sst I; (Xh) Xho I.

The demonstrated lack of involvement of c-*abl* and *bcr* sequences (2,21) in some Ph'-negative CML patients suggests that in these cases the disease may have a different molecular basis. Indeed, clinical differences exist between cases of Ph'-positive and Ph'-negative CML. In a study of 25 cases of Ph'-negative CML, all but one were found to represent myeloproliferative conditions and myelodysplastic syndromes (36) and not CML. These data confirm and strengthen the hypothesis that the genetic rearrangement of *bcr* and the *abl* oncogene is a characteristic of true CML and would argue that the term *CML* should be restricted to those patients with such rearrangements in their DNA.

The *bcr* is Part of a Gene

It is obvious that the Philadelphia translocation has a profound influence on the cellular c-*abl* proto-oncogene, resulting in the disruption of the linkage between the gene and sequences 5' to it. We next addressed the question of how the sequences on chromosome 22 were affected by the chromosomal break. One possibility was that *bcr* could be a gene or part of a gene that would be disrupted as a consequence of the Ph' translocation.

To examine whether *bcr* contains protein-encoding regions, probes from *bcr* were tested for their ability to hybridize to cDNA sequences. One of the probes tested, a 0.6-kb Hind III/Bam HI *bcr* restriction-enzyme fragment from the 5' of *bcr* (Fig. 3A), hybridized to sequences in a human fibroblast cDNA library and was subsequently used for the isolation of several cDNA clones.

The largest cDNA, V1-3, containing an insert of 2.2 kb, was characterized in detail by restriction-enzyme mapping and sequence analysis (24). A long, open reading frame could be identified, having a capacity to code for 589 amino acid residues; this would correspond to a protein of approximately 65,000 daltons. At the 3' end, a polyadenylation signal at nucleotide 2182, followed by a polyA tail beginning at base 2208, indicates that the cDNA contains the complete 3' end of the gene. Although translational start sequences are encountered at the 5' end, the isolated cDNA does not represent a full-length copy mRNA: using probes isolated from V1-3, we have detected the presence of mRNAs with approximate sizes of 4.0 and 6.5 kb (see also below).

The PIR FASTP computer program was used to examine whether any homology could be detected between *phl* and oncogenes, growth factors, or other previously isolated proteins; however, no significant homology was found, indicating that at present, the *phl* protein sequence yields no clues as to its cellular function and its possible role in CML. This gene, which we have previously referred to as the *bcr* gene, has been designated *phl* because of its involvement with the Philadelphia translocation.

The orientation of the *phl* gene on chromosome 22 could be determined by preparing 5' and 3' probes from the V1-3 cDNA, followed by hybridization to cosmids containing human chromosome 22 sequences: The 5' end of the *phl* gene is toward the centromere of chromosome 22 and remains on the Ph' chromosome after the Ph' translocation; the 3' end of the *phl* gene points in the direction of the telomere and is translocated to chromosome 9 in the t(9;22).

Next, we determined the exact position and number of exons within *bcr*: All regions in *bcr* hybridizing to the cDNA V1-3 were sequenced and compared with V1-3. As shown in Fig. 4, five relatively small exons, designated 1 to 5, are present within *bcr* and vary in size from 76 to 105 base pairs (bp); at the 5' of *bcr*, we have identified a minimum of eight additional exons, whereas 3' to it, at least five exons are present (not shown). Since *bcr* was defined as the region on chromosome 22 in which the Ph' breakpoints are found, it follows that the breakpoints occur within a gene. As determined by cDNA hybridizing genomic sequences, the *phl* gene encompasses a region of 67 kb at minimum and contains a minimum of 18 exons.

Once the number and the position of the exons within *bcr* were known, the breakpoints in CML could be redefined as to whether they occur in exon or intron regions. For some CML DNAs, this assignment is easily done. In previous studies, we have demonstrated a breakpoint within a 1.2-kb HBg *bcr* fragment (Fig. 3A) in some of the CML DNAs such as that of patient C481 (21). Since no coding sequences are located within this region (Fig. 4), patients such as C481 must have a chromosomal breakpoint in the intron between the exons designated 3 and 4.

The exact location of the breakpoints in the DNAs of patients 0311068 and 7701C could not be determined from genomic Southern blot analysis. Therefore, $9q^+$ breakpoint fragments from these DNAs were cloned (data not shown), and restriction enzyme analysis followed by Southern hybridization enabled us to lo-

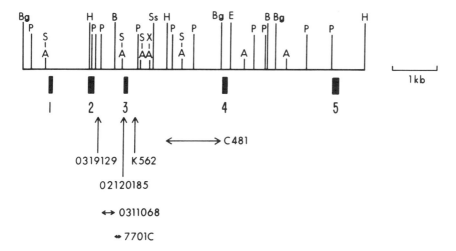

Fig. 4. Philadelphia chromosomal breakpoints within the *bcr*. A detailed restriction enzyme map is shown in the *upper part*; the exons present within *bcr* are indicated with *black boxes* beneath the map and are numbered 1–5. The exact breakpoints in three DNAs are indicated with a *vertical arrow*; for three other DNAs, the breakpoint region was determined by restriction enzyme mapping: (A) Ava I; (B) BamH I; (Bg) Bgl II; (E) EcoR I; (H) Hind III; (P) Pst I; (S) Sma I; (Ss) Sst I; (X) Xho I.

cate the breakpoints between exons 2 and 3 (Fig. 4). On DNA sequencing, the breakpoints in the previously cloned (21,25) $9q^+$ breakpoint fragments of patients 0319129 and 02120185 were also shown to be located between exons 2 and 3. In addition, we cloned the $22q^-$ breakpoint fragment of the cell like K562 (see below). DNA sequence analysis of the breakpoint regions of these fragments, in addition to that of the sequence of the corresponding chromosome 22 regions, allowed us to define the point of translocation for these DNAs (Fig. 4); of 6 breakpoints analyzed in detail, none were found within an exon, indicating that in the Ph' translocation, breakpoints occur within introns of the *phl* gene. Although the exact point of translocation on chromosome 22 differs for each CML DNA, there seems to be little variation in the amount of *phl* exons remaining on the Ph'-chromosome: Four DNAs retain exon 2 and all sequences 5' to it, whereas the DNAs of two CML patients were found to include exon 3 of the *phl* gene.

CML CELL LINE K562 CONTAINS A BREAKPOINT WITHIN *bcr*

In 1975, C.B. Lozzio and B.B. Lozzio (32) reported the isolation of a cell line, K562, from the pleural fluid of an adult patient with CML. Because the cell line expresses phenotypic markers of erythroid lineage, it has been termed erythroleukemic (27).

Heisterkamp et al. (24) (8,42) have shown that the c-*abl* oncogene and a segment of chromosome 22, including the immunoglobulin light-chain constant region in band $q11$ (Cλ) but not extending to the c-*sis* oncogene in band $q12.1$–$q13.3$ (1), are amplified at least fourfold in this cell line. These data suggest that in K562 at least four Ph' chromosomes can be found; however, no Ph' chromosome is visible cytogenetically. Such findings leave unresolved the question of whether amplification of the c-*abl* oncogene in K562 is correlated with the presence of an amplified Ph' translocation. If such a correlation exists, it would unambiguously demonstrate that K562 was derived from the Ph'-positive leukemic cells of the CML patient.

The 1.2-kb HBg probe alone has proved suitable for demonstrating the presence of a breakpoint on chromosome 22 in the majority of all Ph'-positive CML DNAs; for example, abnormal EcoRI restriction enzyme fragments (breakpoint fragments) are clearly present in the DNAs of CML patients 02120185, 0319129, and 0311068 (Fig. 5A). In the K562 cell line, abnormal fragments could not be detected with either EcoRI (Fig. 5A, lane 4) or each of several other restriction enzymes tested (not shown) after hybridization to the 1.2-kb HBg probe. This could indicate that K562 does not contain a breakpoint on chromosome 22 within *bcr*. To examine this more thoroughly, a probe more to the 5' (0.6-kb HB, Fig. 3A) within *bcr* was prepared and hybridized to the DNA of K562 digested with different enzymes. As shown in Fig. 5B, in addition to the normal 5.0-kb Bgl II fragment (lanes 3 and 4), this probe detects abnormal Bgl II fragments in K562 DNA (lane 2). Moreover, one of these fragments is amplified at least fourfold. Since the 0.6-kb HB probe has detected $22q^-$ fragments in the DNAs of a number of CML patients, it seemed likely that the abnormal amplified fragments in K562 represent

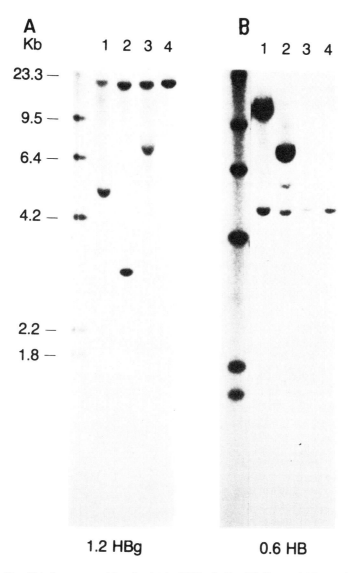

Fig. 5. The Ph' chromosomal breakpoint in K562. **A**: Eco RI digest of 10 μg of DNA from CML patient 02120185 (lane 1), 0319129 (lane 2), 0311068 (lane 3) and from the cell line K562 (lane 4). **B**: K562 DNA digested with Sst I (lane 1) and Bgl II (lane 2); DNAs of human cell lines AG 1732 (lane 3) and AG 2655 (lane 4) digested with Bgl II. **A** and **B** were hybridized with various molecular probes, as shown at the bottom of the figure; the origin of the probes is as indicated in Fig. 3; ^{32}P-labeled Hind III is included in the left lane of each panel as a molecular weight marker.

amplified sequences on the $22q^-$ chromosome. To analyze this in more detail, a cosmid library was constructed from K562 DNA partially digested with Mbo I, according to published procedures (19). Numerous colonies of the approximately 100,000 recombinants hybridized with the 0.6-kb HB probe; three such positives containing overlapping portions of the same region were selected for further restriction enzyme analysis (Fig. 3D). It is evident from a comparison of the detailed restriction enzyme maps of normal chromosome 22 sequences (Fig. 3B) and K562 DNA (Fig. 3C) that the homology between the two terminates 3' to the most 5' Ava I site.

A 1.0 E probe prepared from K562 DNA immediately 3' to the breakpoint (Fig. 3D) hybridizes to DNA isolated from somatic cell hybrids containing human chromosome 9 in the absence of chromosome 22 but not with DNA from hybrids containing chromosome 22 (data not shown). This indicates that the sequences isolated from K562 DNA are chimeric and contain the breakpoint of the $22q$ chromosome. The entire region is amplified at least fourfold; the chromosome 9 specific sequences are also amplified, as demonstrated by the strong hybridization of the 1.0 E probe to K562 DNA in comparison with control DNA (data not shown). Thus, the amplification of chromosome 9 sequences begins at the point where the breakpoint has occurred on chromosome 9 in the Ph' translocation and extends in the direction of the telomere of the chromosome, including the c-*abl* oncogene. The amplified region may be relatively large; the distance analyzed and cloned to date between the breakpoint on chromosome 9 and the most 5' identified c-*abl* exon is, at minimum, 90 kb (J. Groffen, *unpublished data*). Nonetheless, the results indicated above show that the cell line may serve as a model system for Ph'-positive CML. It has been used extensively for this purpose and seems to have retained all *bcr*/c-*abl* characteristics found in patient material.

Analysis of CML Breakpoints by DNA Sequencing

In the Ph' translocation, an illegitimate recombination event occurs between sequences on chromosome 9 and 22. Sequence analysis and hybridization have indicated that the *phl* gene has no homology to previously identified proteins, including the c-*abl* oncogene (24). To investigate the question of why recombination occurs between introns of these genes, the DNA sequence was determined of the breakpoint regions of two CML DNAs and of K562 DNA. As shown in Fig. 6, normal chromosome 22 DNA sequences are identical to those of the K562 breakpoint fragment up to a point approximately in the center of the region shown; 3' of this point, chromosome 9 sequences are joined to those of chromosome 22, and there is no apparent homology between chromosome 9 and 22 sequences in the breakpoint region. Similarly, no homology could be found between chromosome 9 and 22 sequences in the other two CML DNAs, indicating that the translocation cannot be caused by illegitimate recombination between homologous sequences in introns.

```
chromosome 22         CTGGCCGCTGTGGAGTGTTTGTGCTGGTTGATGC
22q⁻ in K562          CTGGCCGCTGTGGAGTGGGTTTTATCAGCTTCCA
                      chromosome 22 ←——————→ chromosome 9
```

Fig. 6. Sequence analysis of the breakpoint of K562.

These results are in concordance with the results obtained from sequence analysis of the t(8;14) translocation in Burkitt's lymphoma, in which no obvious homology was detected between recombined sequences on chromosomes 8 and 14 (4). It is noteworthy, however, that Alu repetitive sequences seem to be involved: On chromosome 9, breakpoints fall within similar types of Alu sequences in the two CML patients. In addition, the break on chromosome 22 of one patient falls within an Alu repeat, and Alu-like sequences are also found in the breakpoint region of K562. Since illegitimate recombination within Alu sequences has been reported in four independent cases of thalassemia and in one case of hypercholesterolemia (30), it is not improbable that Alu repetitive sequences may be hot spots of recombination and as such play a role in the generation of the Ph' translocation.

The Product of the Ph' Translocation Is a Chimeric *bcr/c-abl* mRNA

The results described above indicate that the Ph' translocation results in the fusion of two genes in a head-to-tail fashion: The *phl* gene at the 5' has lost its 3' sequences, which are replaced by c-*abl* sequences from chromosome 9. To examine the effect of the translocation on the expression of these genes, poly A RNA was isolated from K562 and control HELA cells. As shown in Fig. 7, a probe isolated from the most 5' v-*abl* homologous exon of human c-*abl* detects mRNAs of 8.5, 7.0, and 6.0 kb in K562. The 7.0- and 6.0-kb mRNAs were also detected in RNA isolated from HELA cells (data not shown) and represent transcripts from the normal, unrearranged c-*abl* allele on chromosome 9. The 8.5-kb c-*abl* mRNA, however, is found only in K562 and not in non-CML cells, in concordance with results obtained by others (9,15). The level of expression of this 8.5-kb mRNA is significantly higher than that of either the 6.0- or 7.0-kb species and is directly correlated to the presence of the amplified remnants of the Ph' chromosome that include the *phl* gene and the c-*abl* oncogene. This abnormally sized c-*abl* hybridizing mRNA seems to be characteristic for Ph'-positive CML cells: A mRNA of approximately 8.5 kb was also detected in RNA isolated from the leukemic cells of CML patients (6). These data, in combination with the knowledge that the *phl* gene is truncated by the Ph' translocation, resulted in the following model: As a consequence of the Ph' translocation, *bcr* and *abl* genomic sequences are fused into a chimeric gene that could be transcribed into a chimeric *phl/abl* mRNA. To test this hypothesis, two probes were isolated from the *phl* cDNA: probe A corres-

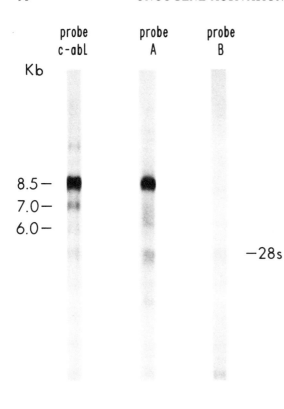

Fig. 7. Hybridization of K562 Northern blot with c-*abl* and *phl*-probes. PolyA$^+$ RNA (20 μg) of K562 was run in the presence of formaldehyde on a 1% agarose gel and transferred to nitrocellulose. For probes we used a 0.6-kb EcoR I/BamH I c-*abl* fragment, a 5' *phl* (probe A) fragment, and a 3' *phl* (probe B) fragment (isolated from the *phl* cDNA V1-3) cloned into pSP vectors and labeled using SP6 RNA polymerase.

ponding to the 5' region of the cDNA ending in exon 2 (Fig. 4) and probe B starting in exon 5 and encompassing 3' cDNA sequences.

If the 8.5-kb *abl* hybridizing mRNA is chimeric, it should hybridize to probe A, which contains *phl* exons 5' to the breakpoint; however, the mRNA should not hybridize to probe B, since 3' exons of *phl* have been translocated to chromosome 9. This hypothesis proved to be correct, because A was found to hybridize with the 8.5-kb mRNA, whereas no hybridization could be detected with probe B (Fig. 7). These results unambiguously show that the abnormal 8.5-kb c-*abl* mRNA as found in the CML cell line K562 is chimeric and represents 5' *phl* sequences fused to c-*abl* sequences.

The structure of the chimeric *phl*/c-*abl* mRNA of K562 could be examined in more detail by the molecular cloning of a cDNA copy of it; as predicted from the genomic DNA organization, it encompasses all *phl* exons to the 5' including exon 3 (Fig. 4), and the *phl* sequence is then joined to that of c-*abl* exon A (Fig. 2) in such a fashion that it is in frame (44). Although one c-*abl* exon at minimum is situated 5' to exon A in the genomic DNA, it has apparently been deleted by splicing from the mature chimeric mRNA; the CML cell line EM2 has a similarly structured chimeric mRNA, with *phl* exon 3 joined to c-*abl* exon A (44). Although most Ph'-positive CML cells seem to contain a similarly sized abnormal transcript of 8 to 8.5 kb [all of 25 patients (16); all of 5 patients (J. Groffen et al., *unpub-*

lished observations)], two cases with an additional transcript of 9.0 kb have been reported (16). This indicates that in some cases additional *phl* and/or c-*abl* exons may be included in the transcript.

The Chimeric mRNA is Translated

Using v-*abl* antisera, the normal cellular product of the human c-*abl* oncogene has been identified as a phosphoprotein with a molecular weight of 145,000 daltons (P145). Although the P145 is phosphorylated *in vivo* on serine, it does not exhibit tyrosine phosphorylation activity *in vivo* or *in vitro* (35). In contrast, v-*abl* is a tyrosine-specific protein kinase *in vitro* (48) and is labeled on tyrosine *in vivo* (41); the enzymatic activity of v-*abl* is strongly correlated with its transforming ability (39). In K562, v-*abl* antisera precipitate two abnormally sized proteins of 210,000 and 190,000 molecular weight (P210 and P190) in addition to P145 (29). Although P145 exhibits no kinase activity, P210 and P190 are phosphorylated *in vivo* and *in vitro* on tyrosine. These abnormal proteins resemble v-*abl* closely in that they all have a similar *cis*- and *trans*- acting phosphorylation activity and a similar extent of inhibition by certain v-*abl* antisera. One major difference between v-*abl* and P210/P190 was found to be that the former is not phosphorylated on serine, whereas the latter is labeled significantly on serine in an *in vitro* kinase reaction (10); possibly, this serine-kinase activity could be present on a separate protein that is tightly associated with P210 and is thus immunoprecipitated by v-*abl* antisera.

Abnormally sized proteins of an approximate molecular weight of 210 that could be immunoprecipitated by v-*abl* antisera have also been detected in phosphate-labeled CML cell lines EM2 and BV173; as the P210 in K562, these proteins exhibited an associated tyrosine kinase activity *in vitro*. In addition, a P210 was identified in ^{32}P-labeled cells of three independent CML patients (28). These results indicate that the chimeric *phl*/c-*abl* mRNA present in Ph'-positive CML cell lines and patient material is translated into a chimeric protein (Fig. 8) and that the difference in molecular weight between P210 and the normal c-*abl* P145 is caused in part by the addition of a *phl* moiety to the amino terminus of P145.

SUMMARY AND DISCUSSION

The results in this chapter describe an increasing insight into the molecular events occurring as a consequence of the first translocation identified as specifically associated with one type of leukemia: the Philadelphia translocation in CML. In this translocation, chromosomal breaks occur within one gene on chromosome 22 and within or 5' to a second gene on chromosome 9; a reciprocal exchange takes place between the segments 22 $q11$-$q$ter and 9q34-$q$ter, and the resulting recombinant chromosome 22 (the 22q^- chromosome) contains a hybrid gene, consisting of 5' regulatory, promotor, and exon sequences of the *phl* gene on chromo-

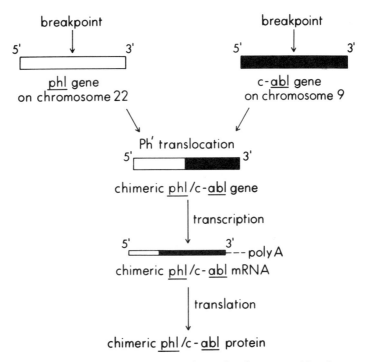

Fig. 8. Chimeric mRNA in CML. The top line shows the chromosomal breaks occurring on chromosomes 22 and 9. The following line depicts the situation on the Ph' chromosome in CML, where chromosome 9 sequences (c-*abl*) are linked to chromosomal 22 sequences within *bcr*. This chimeric gene is transcribed into a chimeric 8.5-kb mRNA consisting of 5' *phl* and 3' c-*abl* sequences that are translated into protein.

some 22 fused to 3' exons and polyadenylation/termination sequences of the c-*abl* oncogene from chromosome 9. It is therefore not unlikely that such a hybrid gene will be responsive to putative transcriptional regulatory signals directed at the normal *phl* gene. Data are available that possibly indicate that the normal *phl* gene and the chimeric one are regulated differently; in CML bone marrow cells, the chimeric *phl*/c-*abl* mRNA is expressed at a higher level than the normal 6.5- and 4.0-kb *phl* mRNAs (J. Groffen et al., *unpublished observations*). Although the chimeric *phl*/c-*abl* mRNA is likewise expressed at a much higher level than the normal *phl* transcript in the CML cell line K562 (44; J. Groffen et al., *unpublished observations*), the difference of four- to five-fold between the gene-copy number of the chimeric and the normal gene may account for this difference. The normal cellular function of the *phl* gene product is unknown; sequence analysis of a *phl* cDNA of 2.2 kb did not reveal homology to previously identified genes (24). The *phl* gene is large (more than 67 kb) and has an elaborate intron-exon organization (a minimum of 18 exons). Despite an extensive search, expression of *phl* mRNA of 4.0 and 6.5 kb was found only in low levels in several SV40 transformed human cell lines, myeloid and lymphoid cell lines, erythroid precursor cells, skin, gut, kidney, and spleen (J. Groffen et al., *unpublished observations*); however, a high level of *phl* expression could possibly be found in certain specialized types of cells not examined by us. For example, a relatively high level of *bcr*-homologous mRNA was found by others (44) in the cell line SMS-SB, which was derived from the leukemic lymphoblasts of a patient with an unusual type of pre-B acute lymphoblastic leukemia (34).

Although relatively more is known about *abl*, its normal cellular function has not been established. The v-*abl* oncogene can be divided into two distinct regions, of which the N-terminal 1.2 kb encodes the tyrosine-specific protein kinase (47). Human c-*abl* contains exons homologous to and colinear with this domain of v-*abl*: The tyrosine phosphorylation acceptor site sequence of v-*abl* is also found within the human c-*abl* oncogene (Fig. 2, region IV, tyr) and is identical in amino acid sequences (20). In addition, the exon designated A in Fig. 2, which is non-v-*abl* homologous, contains sequence homology to v-*src* (J. Groffen et al., *unpublished observations*). Although the normal c-*abl* protein, P145, has no detectable tyrosine kinase activity, the sequence of v-*abl* indicates that it is a member of a family of tyrosine-specific protein kinases, including v-*src*, v-*yes*, v-*fes/fps*, v-*fgr*, v-*erb*-B, v-*ros*, and v-*fms* (for a review, see ref. 5). Since the insulin receptor has a significant degree of homology to v-*ros* and the epidermal growth factor receptor is highly homologous to v-*erb*-B (12,45), it is not unlikely that other proto-oncogenes with tyrosine-specific protein kinase activity may be cellular receptors. Additionally, the *fms* oncogene was found to be the gene encoding the CSF-receptor-I (43). It has been reported that the v-*abl* protein is unlikely to be a transmembrane receptor, since it does not appear to span the plasma membrane.

All breakpoints identified in different patients thus far are 5' of the most 5' v-*abl* homologous exon. Although the amount of different cases examined is small, one would suppose that a minimum exists for the amount of c-*abl* exons present in P210: in analogy to v-*abl*, it would be expected that those exons in c-*abl* encoding the tyrosine kinase activity should be present; however, many more cases should be examined to test this hypothesis vigorously.

In *bcr*, there seems little variation in the amount of *phl* exons included in the chimeric transcript: all chimeric genes analyzed thus far would be transcribed into mRNAs, differing only in the presence or absence of one *phl* exon (Fig. 4) encoding 25 amino acid residues. Since we do not know what conformational changes have been induced by the addition of the *phl* polypeptide to the N-terminus of c-*abl* P145, the functional significance, if any, of the presence or absence of one extra *phl* exon is unclear. However, all evidence available does point to one conclusion: that the attachment of a *phl* moiety to the c-*abl* protein has unmasked associated tyrosine kinase activity. In analogy to the tyrosine kinase activity of v-*abl*, it is not unlikely that this enzymatic activity in P210 may be involved in cellular transformation.

Circumstantial evidence indicates that, at minimum, the production of an active P210 must confer some selective advantage on cells expressing it; our results indicate no obvious reason (such as illegitimate homologous recombination) for the t(9;22) to occur. By lack of contrary evidence, we would assume that many rearrangements may take place in cells that are genetically unstable; although random recombinations between c-*abl* and the *phl* gene would then occur in all introns and exons (with the presence of Alu repeats and the larger size of introns favoring recombination in introns), only certain rare configurations, allowing the production of a functional protein, would lead to a slight selective advantage and slow outgrowth of cells carrying them.

Chronic myelocytic leukemia is most likely the result of the monoclonal expansion of a single transformed pluripotent stem cell (14,33). The disease is characterized by an initially mild course extending over a period of time. It has been estimated that the lapse of time between the primary transforming event(s) and the presentation of overt CML may be as long as 8 years (26). Although the "chronic" phase may be relatively benign, the disease invariably progresses into an accelerated phase and "blast crisis," characterized by an increasing amount of circulating undifferentiated blast cells; usually, the patient dies in this phase.

As stated above, the vast majority (more than 95%) of CML patients have the Ph' chromosome in their leukemic cells. Apart from this cytogenetic abnormality, the differences between CML cells and normal bone marrow cells are subtle. Although many proliferative parameters are similar for chronic phase CML cells and normal hematopoietic cells (18), CML cells differ from normal stem cells in that they do not cease proliferation after reaching a certain cell density: In due course, the leukemic Ph'-positive cells will slowly replace the normal cells of the myeloid compartment (7).

Is then the translocation that gives rise to the Ph' chromosome (one of) the primary transforming event(s) in CML and is it directly correlated to the lack of control of proliferation exhibited by Ph'-positive stem cells? Experiments aimed at positively and unambiguously answering the first question are not easily designed, although experiments aimed at answering the second question should now be feasible. Since tumorigenesis is generally regarded as a multistep process, one would assume that CML also progresses in different stages: At minimum, the blast crisis would represent a step involving further aberrations (often accompanied by additional chromosomal abnormalities) in an already abnormal clone of Ph'-positive cells. Fialkow (13) has postulated that the pasthogenesis of CML involves at least three steps, the first of which would be the clonal proliferation of a pluripotent Ph'-negative stem cell. The acquisition of the Ph' chromosome would be a secondary event; this model is based on the occurrence of Ph'-negative cells carrying the G6PD phenotype also found in the original leukemic clone. In addition, other evidence supporting the concept of a primary transformed Ph'1-negative CML cell is available (for a summary, see ref 16).

The identification of the two genes, *abl* and *phl*, directly involved in and affected by the Ph' translocation should enable us to begin examining the significance of the translocation in CML: If the translocation is one of the obligatory transforming events in CML, the chimeric *phl*/c-*abl* gene product should be directly involved in maintenance of the disease. The availability of molecular probes of the critical segment of the Ph' chromosome should facilitate the distinction between Ph'-positive and truly Ph'-negative cells: The lack of a cytogenetically visible Ph'-chromosome does not prove that bcr and c-abl are still in "germline configuration." On isolation of a complete phl/c-abl cDNA, this DNA can be transfected to cell lines and perhaps introduced into cultured primary bone marrow cells; in this way, the transforming potential of the chimeric phl/c-abl protein can be studied. In addition, one can study the effects of the loss of the chimeric phl/c-abl protein from CML cells: The chimeric mRNA present in these cells can be inactivated by transfection with antisense phl RNA; such RNA will prevent the translation of the chimeric phl/c-abl mRNA by forming a hybrid with it. Apart from giving insight into the effects of the chimeric phl/c-abl protein, analogous experiments may shed light on the function of the normal phl and the normal c-abl gene products.

ACKNOWLEDGMENTS

We thank everyone who has contributed to this research. In addition, we thank Sue Marcus for typing of the manuscript and J.R. Stephenson for helpful comments.

REFERENCES

1. Bartram, C.R., de Klein, A., Hagemeijer, A., Grosveld, G., Heisterkamp, H., and Groffen, J. (1984): Localization of the human c-sis oncogene in Ph'-positive and Ph'-negative chronic myelocytic leukemia by in situ hybridization. Blood, 63:223–225.
2. Bartram, C.R., de Klein, A., Hagemeijer, A., van Agthoven, T., van Kessel, A.G., Bootsma, D., Grosveld, G., Ferguson-Smith, M.A., Davies, T., Stone, M., Heisterkamp, N., Stephenson, J.R., and Groffen, J. (1983): Translocation of c-abl oncogene correlates with the presence of a Philadelphia chromosome in chronic myelocytic leukaemia. Nature, 306:277–280.
3. Bartram, C.R., Kleihauer, E., de Klein, A., Grosveld, G., Teyssier, J.J., Heisterkamp, N., and Groffen, J. (1985): c-abl and bcr Are rearranged in a Ph'-negative CML patient. EMBO J., 4:683–686.
4. Battey, J., Moulding, C., Taub, R., Murphy, W., Stewart, T., Potter, H., Lenoir, G., and Leder, P. (1983): The human c-myc oncogene: structural consequences of translocation into the IgH locus in Burkitt lymphoma. Cell, 34:779–787.
5. Bishop, J.M. (1985): Viral oncogenes. Cell, 42:23–38.
6. Canaani, E., Gale, R.P., and Steiner-Saltz, D. (1984): Altered transcription of an oncogene in chronic myeloid leukaemia. Lancet, 593–595.
7. Clarkson, B., and Rubinow, S.I. (1977): Growth kinetics in human leukemia and biochemical regulations of normal and malignant cells. Edited by Drewinko, B., and Humphrey, R., pp. 591–628. The Williams and Wilkins Co., Baltimore.
8. Collins, S.J., and Groudine, M.T. (1983): Rearrangement and amplification of c-abl sequences in the human chronic myelogenous leukemia cell line K562. Proc. Natl. Acad. Sci. U.S.A., 80:4813–4817.
9. Collins, S.J., Kubonishi, I., Miyoshi, I., and Groudine, M.T. (1984):l Altered transcription of the c-abl oncogene in K-562 and other chronic myelogenous leukemia cells. Science, 225:72–74.

10. Davis, R.L., Konopka, J.B., and Witte, O.N. (1985): Activation of the c-*abl* oncogene by viral transduction or chromosomal translocation generates altered c-*abl* proteins with similar *in vitro* kinase properties. *Mol. Cell. Biol.*, 5:204–213.
11. de Klein. A., Geurts van Kessel, A., Grosveld, G., Bartram, C.R., Hagemeijer, A., Bootsma, D., Spurr, N.K., Heisterkamp, N., Groffen, J., and Stephenson, J.R. (1982): A cellular oncogene is translocated to the Philadelphia chromosome in chronic myelocytic leukemia. *Nature*, 300:765–767.
12. Downward, J., Yarden, Y., Mayes, E., Scrace, G., Totty, N., Stockwell, P., Ulrich, A., Schlessinger, J., and Waterfield, M.D. (1984): Close similarity of epidermal growth factor receptor and v-*erb*-B oncogene protein sequences. *Nature*, 307:521–527.
13. Fialkow, P.J. (1984): Clonal evolution of human myeloid leukemias. In: *Genes and Cancer*, edited by Bishop, J.M., Rowley, J.D., and Greaves, M., pp. 2115–226. Alan R. Liss, Inc., New York.
14. Fialkow, P.J., Gartler, S.M., and Yoshida, A. (1967): Clonal origin of chronic myelocytic leukemia in man. *Proc. Natl. Acad. Sci. U.S.A.*, 58:1468–1471.
15. Gale, R.P., and Cannani, E. (1984): A novel 8 kb *abl* RNA transcript in chronic myelogenous leukemia. *Proc. Natl. Acad. Sci. U.S.A.*, 81:5648–5652.
16. Gale, R.P., and Cannani, E. (1985): The molecular biology of chronic myelogenous leukaemia. *Br. J. Haematol.*, 60:395–408.
17. Goff, S.P., Gilboa, E., Witte, O.N., and Baltimore, D. (1980): Structure of the Abelson murine leukemia virus genome and the homologous cellular gene: Studies with cloned viral DNA. *Cell*, 22:777–785.
18. Goto, T., Nishikori, M., Arlin, Z., Gee, T., Kempin, S., Burchenal, J., Strife, A., Wisniewski, D., Lambek, C., Little, C., Jhanwar, S., Chaganti, R., and Clarkson, B. (1982): Growth characteristics of leukemic and normal hematopoietic cells in Ph'+ chronic myelogenous leukemia and effects of intensive treatment. *Blood*, 59:793–808.
19. Groffen, J., Heisterkamp, N., Grosveld, F., Van de Ven, W.J.M., and Stephenson, J.R. (1982): Isolation of human oncogene sequences (v-*fes* homolog) from a cosmid library. *Science*, 216:1136–1138.
20. Groffen, J., Heisterkamp, N., Reynolds, F.H., Jr., and Stephenson, J.R. (1983): Homology between phosphotyrosine acceptor site of human c-*abl* and viral oncogene products. *Nature*, 304:364–369.
21. Groffen, J., Stephenson, J.R., Heisterkamp, N., de Klein, A., Bartram, C.R., and Grosveld, G. (1984): Philadelphia chromosomal breakpoints are clustered within a limited region, *bcr*, on chromosome 22. *Cell*, 36:93–99.
22. Heisterkamp, N., Groffen, J., and Stephenson, J.R. (1983): The human v-*abl* cellular homologue. *J. Mol. Appl. Genet.*, 2:57–68.
23. Heisterkamp, N., Groffen, J., Stephenson, J.R., Spurr, N.K., Goodfellow, P.N., Solomon, E., Carrit, B., and Bodmer, W.F. (1982): Chromosomal localization of human cellular homologues of two viral oncogenes. *Nature*, 299:747–749.
24. Heisterkamp, N., Stam, K., Groffen, J., de Klein, A., and Grosveld, G. (1985): Structural organization of the *bcr* gene and its role in the Ph' translocation. *Nature*, 315:758–761.
25. Heisterkamp, N., Stephenson, J.R., Groffen, J., Hansen, P.F., de Klein, A., Bartram, C.R., and Grosveld, G. (1983): Localization of the c-*abl* oncogene adjacent to a translocation breakpoint in chronic myelocytic leukemia. *Nature*, 306:239–242.
26. Kamada, N., and Uchino, H. (1978): Chronologic sequence in appearance of clinical and laboratory findings characteristic of chronic myelocytic leukemia. *Blood*, 51:843–850.
27. Koeffler, H.P., and Golde, D.W. (1980): Human myeloid leukemia cell lines: a review. *Blood*, 56:344–349.
28. Konopka, J.B., Watanabe, S.M., Singer, J.W., Collins, S.J., and Witte, O.N. (1985): Cell lines and clinical isolates derived from Ph'-positive chronic myelogenous leukemia patients express c-*abl* proteins with a common structural alteration. *Proc. Natl. Acad. Sci. U.S.A.*, 82:1810–1814.
29. Konopka, J.B., Watanabe, S.M., and Witte, O.N. (1984): An alteration of the human c-*abl* protein in K562 leukemia cells unmasks associated tyrosine kinase activity. *Cell*, 37:1035–1042.
30. Lehrman, M.A., Schneider, W.J., Sudhof, T.C., Brown, M.S., Goldstein, J.L., and Russell, D.W. (1985): Mutation in LDL receptor: Alu-Alu recombination deletes exons encoding transmembrane and cytoplasmic domains. *Science*, 227:140–145.

31. Leibowitz, D., Schaefer-Rego, K., Popenoe, D.W., Mears, J.G., and Bank, A. (1985): Variable breakpoints on the Philadelphia chromosome in chronic myelogenous leukemia. *Blood*, 66:243–245.
32. Lozzio, C.B., and Lozzio, B.B. (1975): Human chronic myelogenous leukemia cell line with positive Philadelphia chromosome. *Blood*, 45:321–334.
33. Moore, M.A.S., Ekert, H., Fitzgerald, M.D., and Carmichael, A. (1974): Evidence for the clonal origin of chronic myeloid leukemia from a sex chromosome mosaic. Clinical, cytogenetic and marrow culture studies. *Blood*, 43:15–22.
34. Ozanne, B., Wheeler, T., Zack, J., Smith, G., and Dale, B. (1982): Transforming gene of a human leukaemia cell is unrelated to the expressed tumour virus related gene of the cell. *Nature*, 299:744–747.
35. Ponticelli, A.S., Whitlock, C.A., Rosenberg, N., and Witte, O.N. (1982): *In vivo* tyrosine phosphorylations of the Abelson virus transforming protein are absent in its normal cell homolog. *Cell*, 29:953–960.
36. Pugh, W.C., Pearson, M., Vardiman, J.W., and Rowley, J.D. (1985): Philadelphia chromosome-negative chronic myelogenous leukaemia: a morphological reassessment. *Br. J. Haematol.*, 60:457–467.
37. Reddy, E., Smith, M., and Srinivasan, A. (1983): Nucleotide sequence of Abelson murine leukemia virus genome; structural similarity of its transforming gene product to other *onc* gene products with tyrosine-specific kinase activity. *Proc. Natl. Acad. Sci. U.S.A.*, 80:3623–3617.
38. Reddy, E., Smith, M., and Srinivasan, A. (1983): Correction. *Proc. Natl. Acad. Sci. U.S.A.*, 80:7372.
39. Rosenberg, N.E., Clark, D.R., and Witte, O.N. (1980): Abelson murine leukemia virus mutants deficient in kinase activity and lymphoid cell transformation. *J. Virol.*, 36:766–774.
40. Rowley, J.D. (1973): A new consistent chromosomal abnormality in chronic myelogenous leukaemia identified by quinacrine fluorescence and Giemsa staining. *Nature*, 243:290–293.
41. Sefton, B.M., Hunter, T., and Raschke, W.C. (1981): Evidence that the Abelson virus protein functions *in vivo* as a protein kinase that phosphorylate tyrosine. *Proc. Natl. Acad. Sci. U.S.A.*, 78:1552–1556.
42. Selden, J.R., Emanuel, B.S., Wang, E., Cannizzaro, L., Palumbo, A., Erikson, J., Nowell, P.C., Rovera, G., and Croce, C.M. (1983): Amplified Cλ and c-*abl* genes are on the same marker chromosome in K562 leukemia cells. *Proc. Natl. Acad. Sci. U.S.A.*, 80:7289–7292.
43. Sherr, C.J., Rettenmier, C.W., Sacca, R., Roussel, M.F., Look, A.T., and Stanley, E.R. (1985): The c-*fms* proto-oncogene product is related to the receptor for the mononuclear phagocyte growth factor, CSF-1. *Cell*, 41:665–676.
44. Shtivelman, E., Lifshitz, B., Gale, R.P., and Canaani, E. (1985): Fused transcript of *abl* and *bcr* genes in chronic myelogenous leukaemia. *Nature*, 315:550–554.
45. Ullrich, A., Coussens, L., Hayflick, J.S., Dull, T.J., Gray, A., Tam, A.W., Lee, J., Yarden, Y., Libermann, T.A., Schlessinger, J., Downward, J., Mayes, E.L.V., Whittle, N., Waterfield, M.D., and Seeburg, P.H. (1984): Human epidermal growth factor receptor cDNA sequence and aberrant expression of the amplified gene in A431 epidermoid carcinoma cells. *Nature*, 309:418–425.
46. Wang, J.Y.J., and Baltimore, D. (1983): Cellular RNA homologous to the Abelson murine leukemia virus transforming gene: expression and relationship to the viral sequence. *Mol. Cell. Biol.*, 3:773–779.
47. Wang, J.Y.J., Ledley, F., Goff, S., Lee, R., Groner, Y., and Baltimore, D. (1984): The mouse c-*abl* locus: molecular cloning and characterization. *Cell*, 36:349–356.
48. Witte, O., Dasgupta, A., and Baltimore, D. (1980): Abelson murine leukaemia virus protein is phosphorylated *in vitro* to form phosphotyrosine. *Nature*, 283:826–831.

Plasmacytoma Development in BALB/c Mice

Michael Potter

National Cancer Institute, National Institutes of Health, Bethesda, Maryland 20505

Tumor progression is a sequence of genetic changes that begins in a normal cell and terminates with the appearance of an autonomously proliferating destructive cell type. Progression in most tumor systems usually takes place over a relatively long period of time. During the latent period, abnormal cellular proliferations are frequently found, e.g., gastrointestinal polyps (30), mammary gland plaques (24), hepatic foci (23,73), papillomas (37), and the granulocytic proliferation of chronic myelogenous leukemia. There are two histogenetic modes of progression: (a) a plateau form in which succeeding steps arise from precursors in benign proliferations, and (b) a stem cell form in which the progressive changes occur cryptically in stem cells. In the latter form, the abnormal proliferation represents a manifestation but not a step in the process.

The induction of plasmacytomas in genetically susceptible strains of mice provides an experimental model system for the study of tumor progression in a hematopoietic cell lineage and is probably an example of a plateau form of progression. Four cellular stages in this process that are potentially accessible for analysis are (a) the normal immunoglobulin-secreting cell in which the initial changes occur; (b) a recently recognized intermediate proliferating plasma cell that occurs in foci; (c) a condition-dependent primary plasmacytoma cell that requires special conditions for growth; and (d) an autonomous plasmacytoma cell. The precise identity of the normal cell in which plasmacytoma development begins is not established with certainty, but the available evidence implicates a mature cell in the B-lymphocyte lineage.

Three biological features of plasmacytomagenesis in mice that directly relate to a discussion of tumor progression are (a) the physical and chemical properties of the inducing agents; (b) the fact that plasmacytoma development occurs only in genetically susceptible inbred strains and is not a universal response in all inbred strains of mice; and (c) that 98% of all plasmacytomas in BALB/cAnPt and related BALB/c congenic mice have chromosomal anomalies or translocations that occur in the D2/D3 band of chromosome 15.

INDUCING AGENTS AND THEIR MODE OF ACTION

Plasma cell tumors are induced in genetically susceptible strains of mice by two contrasting types of agents; solid plastics and paraffin oils. These agents must be introduced intraperitoneally, and tumor development takes place in the peritoneal connective tissues.

Most of the current work on plasmacytomagenesis is done with pristane (2,6,10,14-tetramethylpentadecane), since it is a chemically defined agent and because it appears to be more potent than the physically similar light-white medicinal mineral oils (57). Pristane, a branched-chain alkane, is probably not metabolized and for this reason cannot be regarded as a direct carcinogen for B-lymphocytes or plasma cells. Instead, the carcinogenic mode of action of pristane is indirect, and its activity appears to be related to the phagocytic and inflammatory processes that are generated in response to the intraperitoneal presence of the oil. This response consists of (a) the migration of macrophages and neutrophils into the peritoneal space, (b) phagocytosis of oil droplets, and (c) formation of an organized fixed tissue (oil granuloma). The phagocytic process in BALB/c mice recruits large numbers of neutrophils and macrophages (7,64) to eliminate the oil. It is assumed that the prolonged presence of free oil in the peritoneum stimulates the bone marrow to produce a continuous supply of these cells.

Solid-plastic materials, such as Millipore diffusion chambers implanted intraperitoneally, were actually the first agents to be described (41; see also ref. 57 for discussion and references). The common pathological change induced by these contrasting agents has not yet been identified, but since plastics do not induce an oil granuloma, it is reasonable to conclude the oil granuloma is not a *sine qua non* for plasmacytomagenesis.

Induction studies have indicated that mice develop plasmacytomas when the dose of pristane is increased (62). Using single injections, it was found that more mice developed plasmacytomas by 365 days with a 1-ml dose than with 0.5 ml (62), i.e., approximately 40% versus 25%, respectively. The most effective dose schedule that is now used routinely is 0.5 ml given on days 0, 60, and 120; this gives a mean incidence of 60% by 365 days (62). These findings suggest that the continued presence of free oil in the peritoneal space is an important factor in plasmacytoma formation.

The intraperitoneal injection of mineral oil or pristane induces two processes. First, the free oil in the peritoneal space directly or indirectly causes monocytes, macrophages, and neutrophils to migrate into this cavity to endocytose oil droplets (7). Second, the prolonged phagocytic process stimulates hematopoietic tissue to produce more cells. Fully developed macrophages as well as more primitive cells in this differentiation series are seen; even occasional mitotic figures can be observed in macrophages. The phagocytes begin to adhere and accumulate on peritoneal surfaces, particularly at the junctions of the mesentery and the intestine, the diaphragm, abdominal wall, and the omental fat surrounding the pancreas. Cells carrying droplets of oil appear to migrate between mesothelial cells where they

ORIGIN OF RADICALS

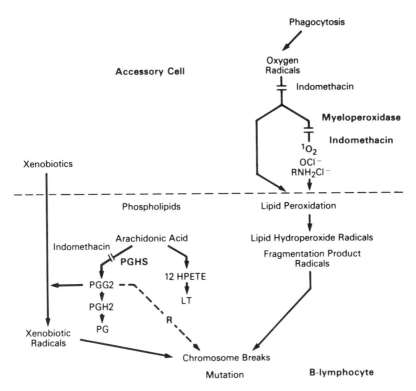

FIG. 1. Scheme showing the origin of oxygen and lipid radicals that potentially induce chromosome breaks in B-lymphocyte chromosomes. Oxygen radicals (e.g., $HO \cdot O^-_2$) are generated during endocytosis in inflammatory phagocytes (accessory cells). In neutrophils, the enzyme myeloperoxidase (MPO), which is packaged in granules, can be released into the extracellular environment. MPO reacts with CL^- and H_2O_2 to produce OCl^- and other radicals. These can interact with lipids in the plasma membrane of B-lymphocytes to produce lipid hydroperoxy radicals or lipid fragmentation products (which are lipid radicals). Also, radicals can be potentially generated intracellularly during the peroxidation of arachidonic acid. This has not yet been directly demonstrated to occur in B-lymphocytes and must be regarded as hypothetical for the case of B-lymphocytes or plasma cells. Sites where indomethacin could inhibit radical production are shown (see text).

adhere to the mesenteric connective tissues (1). The oil granuloma begins forming within a few days, and tissue with its characteristic histology is found in abundance by day 7 after the injection of 0.5 ml pristane (7). The accumulating granulomatous tissue is vascularized by angiogenesis from underlying vessels (1). Some of these angiogenic buds apparently balloon outward to form polyp-like structures containing oil granuloma tissue (1). The organized oil granuloma consists largely of various configurations of macrophages surrounding droplets of oil. These range from signet ring-like single cells engulfing relatively large droplets to groups of macrophages that coalesce on the surface of very large drops. Neutrophils are found scattered throughout. Occasionally, large oil droplets are surrounded by neutrophils. Focal proliferations of myeloid cells (extramedullary myelopoiesis) are also frequently distributed throughout the oil granuloma. The oil granuloma also contains clusters of lymphocytes, some of which are milk spots (1), but others are perivascular accumulations. Plasma cells and lymphocytes are sparsely scattered throughout the oil granuloma. Some plasma cells are in clusters; these cells differ from the foci that contain atypical cells (64) (to be described below) by having very uniform small nuclei. Lymphocytes are also found in the free peritoneal exudate, and large numbers of them can often be seen after day 80 (7,64).

An important feature of the oil granuloma is that emmigrating lymphocytes and plasma cells are exposed to a tissue that is extremely active in phagocytosis and to the products of the phagocytic process, e.g., oxygen radicals. These toxic compounds could attack the membranes of lymphocytes and developing plasma cells, inducing peroxidation of polyunsaturated fatty acids. A chain reaction of organic radical formation that ultimately reaches the DNA of B-lymphocytes and plasma cells could develop (Fig. 1). The clastogenic action of radicals generated by activated inflammatory cells for lymphocytes has been described in other systems (9,18,20) and appears to be applicable to peritoneal plasmacytomagenesis.

GENETIC FACTORS

Plasmacytomas can be induced in approximately 60% of BALB/cAn and other existing BALB/c sublines, with the exception of BALB/cJ by the intraperitoneal injection of paraffin oils or defined alkanes such as pristane (61,62). In contrast to BALB/c, most other commonly used inbred strains (e.g., C57BL, C3H, A/He, DBA/2, AKR, and CBA), are resistant (58). NZB/B1 is the only exception and is susceptible but to a lesser degree than BALB/c (43,45,80). Genes that determine susceptibility and resistance have not yet been identified (58,63).

CHROMOSOMAL TRANSLOCATIONS AND ANOMALIES INVOLVING THE c-*myc* ONCOGENE

An important feature of BALB/c plasmacytomas has been the finding of nonrandom chromosomal breaks and rejoinings that involve the junction between bands D2/D3 in chromosome 15 where the c-*myc* oncogene is located. These have been found in more than 98% of 51 plasmacytomas so far karyotyped (66). The translocations are associated with one or more forms of c-*myc* oncogene deregulation. The transcriptional unit of the c-*myc* oncogene has three exons (E1, E2, and E3) and two introns and spans approximately 3.5 kilobases. It also has a 5' flanking region of undetermined length that contains regulatory sequences (85). Only E2 and E3 contain coding sequences for the *myc* protein, which contains 439 amino acid residues. Because the half-life of normal c-*myc* messenger RNA is approximately 10 min (54), c-*myc* must be transcribed actively to maintain *myc* protein levels in the cell.

The most common translocation, designated rcpt (12;15), is an imprecise reciprocal exchange involving chromosomes 12 and 15 (27). This occurs in approximately 60% of plasmacytomas (58,66) (Fig. 2). The breakpoint sites in chromosome 15 usually occur in a 1.5-kb segment between the beginning of E1 and the end of intron 1 (14). This breakpoint dissociates the coding sequences of c-*myc* E2 and E3 from the centromeric end of the *myc* gene on chromosome 15 and joins them to a site close to the distal end of chromosome 12 (21). The breakpoint in chromosome 12 usually occurs in one of the immunoglobulin heavy-chain switch (S)-region sequences (14). In many cases, the S site used in this recombination has already been rearranged by switching and is therefore an $S_\mu S_\alpha$ or $S_\mu S_\gamma$ complex (14).

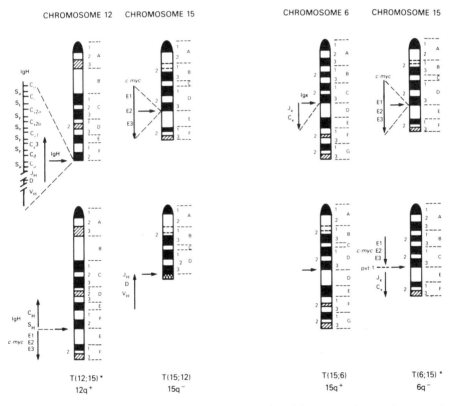

FIG. 2. Two most common reciprocal translocations found in mouse plasmacytomas rcpt (12;15) and rcpt (6;15). The order of genes at the break sites in *myc*-transcribing chromosomes t (12;15) or $12q^+$ (21) and t (6;15) or $6q^-$ (2) are shown. $12q^+$ and $6q^-$ are not active in Ig secretion.

The order of IgH genes in chromosome 12 is centromere-C_H complex-J_H-D-V_H (21) (Fig. 2). Thus, the joining of the distal fragment of chromosome 15 places the sequences of IgC_H and c-*myc* in reverse transcriptional orientation (Fig. 2). Transcripts of c-*myc* are made from t (12;15) ($12q^+$) but not from t (15;12) ($15q^-$) (27) (Fig. 2). The major effect of most (12;15) translocations is to dissociate the regulatory sequences (promoters and other elements located at the 5' end of the gene and including E1 and intron 1) from the coding sequences. Synthesis of *myc* RNA in the absence of its usual promoters requires the activation of "cryptic promoters" in intron 1 and the presence of intron sequences in tumor *myc* mRNA seems to result in long-lived mRNA (54,85). In a second set of rcpt (12;15) translocations, the breakpoint sites on chromosome 15 occur 350 to 500 basepairs (bp) 5' of c-*myc* E1 (85) (Fig. 2). This region probably contains a regulatory element of the c-*myc* oncogene. Breakpoints at this site retain the two normal promoters, but they are utilized in RNA synthesis in an unusual way (85).

The secondmost common nonrandom translocation is rcpt (6;15), which occurs in approximately 16% of BALB/c plasmacytomas (58,66). The breakpoint site on chromosome 6 frequently appears to be in the C_K complex between the J_K4 and C_K (15). Thus, the chromosome 6 breakpoint does not usually involve a rearranging J-region sequence (15). The rcpt (6;15) translocation site on chromosome 15 is within a locus called *pvt*-1 that is located at least 72 kb from c-*myc* (15,81). Using the CAK TEPC 1198 tumor that has an inverse Rb (6;15) translocation [a homologous form of rcpt (6;15) that occurs within a Robertsonian chromosome, which consists of chromosome 6 and chromosome 15, is fused at the centromeres], Banerjee *et al.* (2) have determined that *pvt*-1 is located 3' of c-*myc*. From this it can be inferred that in rcpt (6;15), the t(15;6) ($6q^+$) is the *myc* transcribing chromosome and that the order of genes is centromere C-*myc-pvt*-1 (Fig. 2). C-*myc* transcription is increased in tumors with the rcpt (6;15) translocations (15,46,49), but the molecular basis of this distant position effect is not understood.

Other more unusual rearrangements have been found by sequencing the DNA around (12;15) rearrangement sites (Fig. 3). These rarer cases are associated with transpositions and inversions of chromosome 12 DNA into translocation sites, e.g., as in ABPC45 (22) and PC7183 (78). The J558 tumor has also undergone a second deletional event following the reciprocal exchange (27) (Fig. 3). Several recombinational events must occur to bring about these complex situations.

In typical rcpt (12;15) diploid karyotypes, one normal chromosome 15 and one translocated chromosome 15 are found. In near-tetraploid cells, there are two copies of normal chromosome 15 and two copies of translocated chromosome 15. Further, in rcpt (12;15) the translocated chromosome 12 [i.e., the t(12;15) or $12q^+$-*myc*-transcribing chromosome] is a nonproductively rearranged chromosome (i.e., a chromosome 12 that is not rearranged in a way suitable to synthesize functional Ig heavy-chain mRNA). The same relationship probably also pertains to rcpt (6;15) translocations occurring in *k*-chain producing plasmacytomas, since the rcpt (6;15) breakpoint disrupts the *k*-locus in chromosome 6. An interesting phenomenon is observed in IgA-secreting plasmacytomas with rcpt (12;15); namely, that

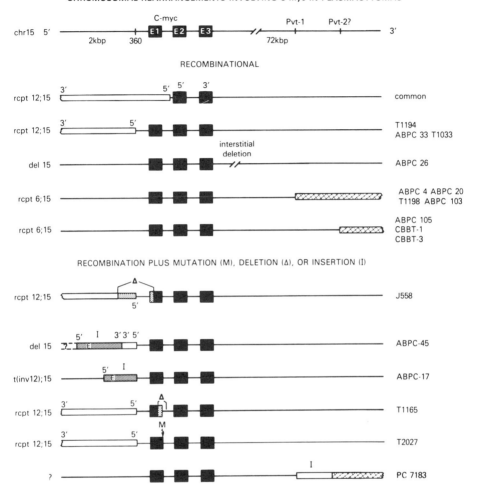

FIG. 3. Summary of chromosomal anomalies in mouse plasmacytomas. The *solid line* indicates the c-*myc* locus in chromosome 15 DNA *(top line)*. E1, E2, E3 are c-*myc* exons marked in the *upper line* but represented by *solid blocks* elsewhere. The *unshaded thick blocks* indicate chromosome 12 DNA in 3' to 5' orientation. The *hatched blocks* in ABPC45 and ABPC17 indicate chromosome 12 DNA in 5' to 3' orientation that contain E (enhancer) sequences. The *cross-hatched blocks* in the *lower three lines* indicate chromosome 6 DNA: (Pvt) plasmacytoma variant translocation; (T) TEPC. In J558 there was a deleted region that is indicated by Δ and by fine *cross-hatching*. The details of the translocational events are given in the following references: common (14); T1194, A33, and T1033 (85); J558 (27); ABPC45 (22); ABPC17 (13); ABPC26 (84); PC7183 (78); ABPC4, ABPC20, T1198, ABPC103 (15); A105, CBBT-1, CBBT-3 (15); T1165, T2027 (85); S. Bauer and J.F. Mushinski *(personal communication)*.

the breakpoint in the translocated chromosome 12 usually involves the switch-α (S_α) region (Table 1). Thus, these cells secrete IgA utilizing the productively rearranged and switched chromosome but have additionally undergone a switch to S_α in the nonproductive chromosome as well (Table 1). This may reflect an intense action of factors that control switching in precursors of plasmacytoma cells.

Although the chromosome breakage in Ig gene loci can be plausibly related to the Ig gene rearrangements and switching processes that normally break and rejoin Ig gene DNA, the underlying basis for the chromosome 15 breakage has no obvious explanation. At present, the chromosome 15 breakpoints (in the 5' flanking regions, in the first intron and exon, in *pvt*-1, and the hypothetical *pvt*-2) are thought to reflect fragile chromosomal sites, which for reasons not yet understood are prone to "spontaneous" breakage (86).

The chromosome 15 abnormalities, though seemingly subtle are generally thought to have profound biological significance in the plasma cell by increasing the availability of c-*myc* transcripts and *myc* gene product throughout the cell cycle. This could be the result of continuous transcription of *myc* throughout the cell cycle or an alteration in the RNA transcript that affects the half-life of this normally short-lived messenger (54). Although the function of the c-*myc* gene product is not known, its presence in the nucleus and its DNA binding activities suggest that it may be a factor required for driving the cell through a segment of the mitotic cycle (29). Initial evidence showed that lymphocyte mitogens induce transcription of the c-*myc* oncogene in B and T lymphocytes and in various cell lines (32,33). Subsequently, it has been found that *myc* gene transcription and *myc* gene

TABLE 1. *IgC$_H$ switch (S) target sites in IgA-secreting plasmacytomas*[a]

Tumor	Karyotype[a]	Break site on chromosome 12	Ref.
J558	(12;15)	$S_\mu S_\alpha$[b]	27
W267	(12;15)	$S_\mu S_\alpha$[c]	27
MOPC315	(12;15)	$S_\mu S_\alpha$	14
TEPC15	(12;15)	5' of C_α	14
McPC603	(12;15)	$S_\mu S_\alpha$	32
S117	(12;15)	$S_\mu S_\alpha$	14
S194	(12;15)	$S_\mu S_\alpha$	14
EPC109	(12;15)	$S_\mu S_\alpha$	14
BFPC61	(12;15)	$S_\mu S_\alpha$	14
S107	(12;15)	$S_\mu S_\alpha$	14
TEPC1165	rcpt (12;15)	5' of C_α	85

[a]The plasmacytoma that has not been karyotyped is indicated by (12;15) and the (12;15) translocation has been determined by DNA sequencing of the break site or by restriction-fragment DNA hybridization.

[b]The original recombination was to an $S_\mu S_\alpha$ fusion complex and was followed by a deletion at the t (12;15) recombination site (27).

[c]The original recombination was to S_α; Gerondakis *et al.* (27) suggest the S_μ sequence may have been spliced in after the translocation.

protein are produced constitutively throughout the mitotic cycle in dividing cell lines (6,29,67,76). Activation of the c-*myc* gene in human B cells, however, does not make these cells competent to complete the mitotic cycle (75). Since the physiological function of c-*myc* product is not known at this time, the effects of changing the levels of c-*myc* transcripts and c-*myc* product at various times in the cell cycle are not understood. However, deregulations of c-*myc* in which high levels of c-*myc* gene product are generated continuously may be inappropriate in differentiated plasma cells that presumably have a high proclivity to enter and remain in a post mitotic Go state. Deregulation of c-*myc* may block the ability of an immunoglobulin-secreting cell to remain in Go, thereby setting the conditions for potential recycling. The fact that all but one BALB/c plasmacytoma has some form of D2/D3 band chromosome 15 anomaly indicates that the cells in which these translocations occurred were at a selective advantage to continue proliferation and development into plasmacytoma cells. An extraordinary feature of these translocations is the variety of modifications that can apparently achieve the end result of increasing c-*myc* transcripts and presumably *myc* protein availability throughout the cell cycle. This may reflect extensive chromosomal recombinations and rejoinings that occur in these cells to produce a few meaningful "mutants." The value of the mutation to the cell in which it occurs must be immediately felt. Further, c-*myc* translocations could create a primary mutation that begins the process of progression. Although it can only be a matter of speculation at this time, the frequency of chromosomal breaks affecting c-*myc* in plasmacytomas indeed suggests that this is a critical lesion. This can only be established when the function of c-*myc* product is determined.

The four different cellular stages thus far defined that are relevant to plasmacytoma progression are now to be discussed.

The Normal Immunoglobulin-Secreting Cell

Immunoglobulin (Ig) secretion begins very soon after appropriate signals have triggered proliferation of the B-lymphocyte and at a time when the cell still has more of the characteristics (cell surface markers, morphology) of a B-lymphocyte than a plasma cell. It has been proposed that much of Ig secretion in normal immune responses is generated not by mature plasma cells but by more primitive cells in the series (26). In this early stage of development, B cells have a proclivity to reenter the circulation and relocate in distant lymphoid tissues as memory cells (35). Plasma cells, however, are usually considered to be the morphological end stage of the Ig-secreting developmental process and are associated with loss of mitotic activity and imminent elimination (72). The striking features of many plasmacytoma cells in mice are an abundant cytoplasm rich in rough endoplasmic reticulum and Golgi vesicles and active secretion of Ig. Some plasmacytomas, however, have characteristics suggestive of plasmablasts with a more lymphocyte-like nucleus and a relatively small amount of cytoplasm. The MOPC 315 plasmacytoma is yet another type and has two cell forms: a proliferating lymphoid stem cell and a mature Ig-secreting cell (69). This range of morphologies suggests that mouse plasmacytomas develop at different stages in the B-lymphoblast → plasmablast → plasma cell transition.

Immunoglobulin-secreting cells and their immediate precursors, the B-lymphocytes, are the likely candidates for the cells in which the first critical changes in the plasmacytoma progression process occur. One of the striking characteristics seen in experimental forms of Ig-secreting tumors is strong heavy-chain isotype predilection. In the BALB/c plasmacytoma system, IgA is the predominant form; in NZB/B1, it is IgG (43,45,80); and in spontaneous rat immunocytomas, the rare IgE class is the major isotype expressed (3). Factors and mechanisms that determine isotype specificity and the selective proliferation of cells with particular isotypes may also play a role in the early and initial steps in neoplastic progression.

Myeloma proteins of the IgA class are produced in more than 60% of transplantable BALB/c plasmacytomas (44,56). When serum from primary cases is examined, the actual number is lower, but this is due to the large number of primary tumors that secrete only light chains or that are nonproducers (44). The preponderance of the IgA class expression in BALB/c plasmacytomas suggests that the subset of cells that have undergone an IgC_H switching process to the switch α S_α are at increased risk in BALB/c mice of progressing to plasmacytoma cells. The importance of the IgA switching process in BALB/c is further supported by the fact that many of the tumors with rcpt (12;15) translocations have a break in the S_α region of chromosome 12. Moreover, in all of the IgA-secreting myelomas so far studied, the rcpt (12;15) translocation site in the nonproductive chromosome is also in S_α (Table 1). This suggests that $S_\mu \rightarrow S_\alpha$ switches that took place on both chromosomes 12 could have occurred at the same time. It is tempting to speculate that the involvement of S_α in the translocated, non-Ig-producing chromosome reflects an overactivity of specific recombinases. Specific recombination factors (switch recombinases) have been proposed by Davis et al. (17) to determine specific kinds of switches, but so far specific recombinases have not been isolated. Cebra et al. (8) have postulated that switching is promoted by repeated cell division. The switching process in plasmacytomas and in some forms of normal differentiation is associated with deletion of intervening parts of the IgC_H complex. Thus, progressive switching could exhaust all possibilities and come to rest at the S_α site. If this is the case, then a specific switch recombinase may not be required, and instead a process that stimulates cell division in B-lymphocytes could be the factor that leads to switching to S_α in both chromosomes 12 and places cells at increased risk for subsequent plasmacytomagenesis.

Two precursor B-cell populations are candidates for the precursors of plasmacytomas, the rapidly dividing B-cells in the germinal centers of Peyer's patches and bone marrow lymphocytes, which may include memory cells or newly differentiated B-lymphocytes. A considerable body of evidence beginning with the original observations of Craig and Cebra (16) supports the hypothesis that a large proportion of B-lymphocytes in Peyer's patches are committed to IgA secretion (71,77) (i.e., they have undergone an $S_\mu \rightarrow S_\alpha$ switch). Immunoglobulin A-programmed B-lymphocytes normally leave Peyer's patches via lymphatics, sojourn in the mesenteric node, and after approximately 7 days enter the lamina propria of the gut where they complete the terminal maturation to the plasma cell stage. Isolated IgA-secreting cells are found in abundance in the oil granulomas of BALB/c mice (E.B. Mushinski and M. Potter, *unpublished observations*), suggesting that the oil granuloma attracts these circulating cells.

A second source of the IgA-programmed cells that arrive in the oil granuloma could be directly from the bone marrow. They could come either from memory cells that had homed back into the marrow (35) or from stem cells. Both kinds of cells could have been stimulated to proliferate by factors emanating from the peritoneum in response to the increased mobilization and utilization of macrophages and neutrophils involved in the phagocytosis of oil. Pietrangeli and Osmond (55) have shown that the intraperitoneal injection of mineral oil increases cell division in bone marrow lymphocytes.

An important question in BALB/c plasmacytomagenesis is the time and site of origin of rcpt (12;15). Does it occur in rapidly dividing cells outside the peritoneum? If so, it would be mediated by special conditions in Peyer's patches or in the bone marrow.

An alternative hypothetical scheme for the time and site of origin of the rcpt (12;15) translocation is based on the possibility that the translocation event is dependent on, but nonetheless temporally and spatially dissociated from, the $S_\mu + S_\alpha$ switching event. For example, an intense switching process could switch both chromosomes 12 to S_α. If so, it could then be proposed that the $S_\mu + S_\alpha$ switch site on the nonproductive chromosome forms a fragile site that is vulnerable to subsequent double-strand breakage. The potential for generating oxygen and lipid radicals during the phagocytosis of oil droplets could provide clastogens that cause double-strand breaks in Ig-secreting cells in the oil granulomas. If this is so, it now becomes possible to consider that the c-*myc* gene rearrangement and deregulation event is dissociated in time and cell (in the B-cell lineage) from the $S_\mu + S_\alpha$ differentiation event. Thus, the rcpt (12;15) could be a late event in plasmacytomagenesis and occur during the plasma cell proliferative stage that takes place in the oil granuloma under a different set of influences than those which control switching. The rcpt (6;15) translocations do not involve IgH switching genes. These translocations occur between the IgK locus usually between J_K and C_K and are not thought to be mediated by $V \rightarrow J$ rearrangement (15). The breaks in chromosome 6, then, probably occur simultaneously with the chromosome 15 breaks to produce rcpt (6;15) and probably not during the Ig light-chain gene-rearrangement process.

Plasma Cell Proliferative Stage

A striking feature of peritoneal plasmacytomagenesis is the appearance of multiple foci of proliferating plasma cells in peritoneal connective tissues and oil granuloma. These are found during the latent period (25–120+ days) of plasmacytoma development (64,65). These proliferating plasma cells reflect a fundamental abnormality in growth regulation, since these are Ig-secreting cells, which paradoxically continue to divide.

Focal Proliferations

When the oil granuloma was systematically examined for the first evidence of plasmacytoma development, focal proliferations of plasma cells were found (Fig. 4). The distinguishing characteristics of plasma cells in foci were their relatively intense uptake of stain, which made them easily discernable in methyl green pyronin-stained sections, and the presence of atypical plasma cells in foci (i.e., cells with larger, more irregular nuclei). For purposes of enumeration, only aggregates containing 50 or more cells were considered to be "foci", however, smaller clusters of very similar cell types were also frequently seen. Isolated plasma cells and clusters of relatively normal staining plasma cells with small round nuclei were commonly found in the oil granuloma. Foci were first found in two of 25 mice 25 days after the injection of pristane and continued to appear in more mice and in greater numbers per mouse thereafter (Fig. 5). Some foci were compact, i.e., composed of aggregates of plasma cells without other intervening cell types. Other foci consisted of looser aggregates of cells and in these some plasma cells infiltrated into the surrounding tissue. In mice with multiple foci, there was morphological evidence of spread by metastases and invasion: The characteristic morphology of cells in separate foci was the same, and numerous clusters of these similar-looking cells could be found dispersed throughout the oil granuloma. It was difficult to distinguish on a morphological basis primary *in situ* focal proliferations of plasma cells from secondary foci that had developed from metastatic or invasive

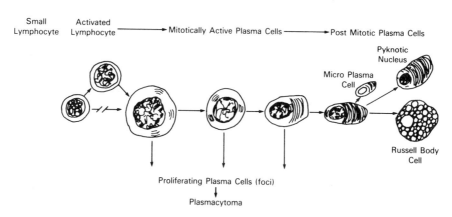

FIG. 4. Scheme of plasma cell development and maturation. The possible origin of foci and plasmacytomas is indicated.

primary cells. Thus, it is not yet clear whether a mouse carrying multiple foci has a multiclonal proliferation or whether a single clonotype has spread extensively at an early time.

Some foci contained small plasma cells with round, uniform nuclei; these were distinguished from normal cells by their hyperchromicity. More atypical plasma cells with large nuclei that had indentations of the nuclear membrane and clumping of the chromatin were seen commonly in foci. The plasma cells in different foci from different mice displayed a continuous variation of cytological appearances ranging from normal to highly atypical forms. The cell sizes within foci also varied from large to small, possibly reflecting the occurrence of polyploidy. As a rule, foci did not have very mature cell types, such as Russell body cells, but a few did, and in these a large number of the cells appeared to be maturing. Although a systematic study has not been completed, the cells in foci at day 50 do not appear to be transplantable *(unpublished observations)*.

The mitotic activity of plasma cells in foci indicates that a critical biochemical lesion has occurred in these cells that allows a highly differentiated cell type that usually is postmitotic (72) to continue proliferating. Proliferation is probably not controlled by the usual signals that act on B-lymphocytes, i.e., antigen and T-helper cells. These factors operate through membrane-bound Ig and other receptors that are probably not functional in Ig-secreting cells. Mitosis is possibly stimulated by other exogenous factors.

Inhibition of Proliferative Foci

The nonsteroidal anti-inflammatory drug indomethacin administered continuously in the drinking water at a concentration of 20 µg/ml after the injection of pristane dramatically inhibits the development of plasmacytomas (64). This treatment does not inhibit the development of the oil granulomatous tissue, the influx of inflammatory cells into the peritoneal space, or the ability of macrophages and neutrophils to phagocytize oil droplets. Strong inhibition of plasmacytoma development can also be observed when indomethacin treatment is delayed until day 60 after pristane injection. Indomethacin treatment neither abrogates the ability of pristane to condition the peritoneum to accept a primary plasmacytoma in transplant nor does it appear to affect the growth of primary plasmacytoma cells in conditioned mice (64). Indomethacin treatment, however, does inhibit the formation of proliferative foci. Further, if indomethacin treatment is delayed until day 60, the number of proliferative foci found at day 100 is appreciably below that observed in controls *(unpublished observations)*. These results indicate that indomethacin slows and inhibits the progression of plasmacytomas and possibly causes regression of some foci.

Indomethacin's primary biological action is as a competitive inhibitor of prostaglandin H synthase (PGHS) (38). PGHS has two enzymatic activities. The first converts arachidonic acid to PGG2 (the cyclooxygenase activity), and the second

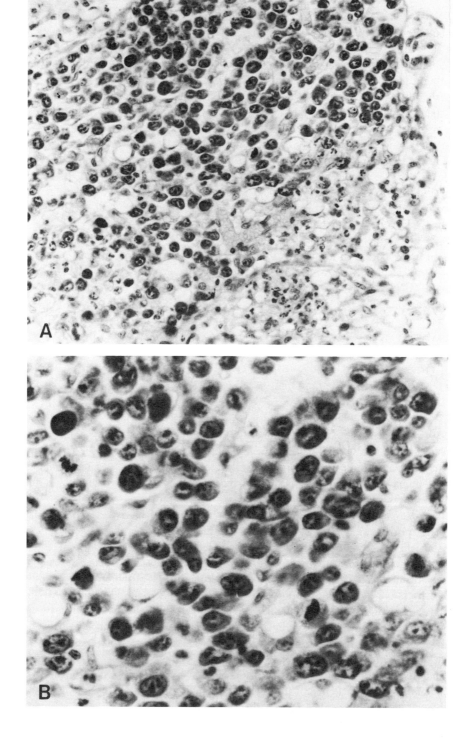

action is a peroxidative reduction of PGG2 to PGH2, the precursor of the physiologically active prostaglandins and thromboxane (Fig. 1). A by-product of the cyclooxygenase reaction is the formation of an intracellular radical (38). Indomethacin blocks the cyclooxygenase reaction (4) and thereby inhibits the formation of radicals as well as prostaglandin biosynthesis. Indomethacin may also be an inhibitor of another heme-containing peroxidase, myeloperoxidase (51,52). Myeloperoxidase is formed in large quantities in neutrophils and is discharged intracellularly into lysozymes, where it generates a bacteriocidal radical HOC1 and singlet oxygen (34). Granules containing myeloperoxidase are also released into the extracellular space, where they can generate radicals in the extracellular spaces.

The relevant cellular target of indomethacin in the oil granuloma is not established with certainty. Two candidates are (a) non-B-lymphocyte accessory cells, i.e., neutrophils or macrophages, and (b) B-lymphocytes or plasma cells. There is evidence in other systems associated with mutagenesis (9,20) and neoplastic transformation (82) that the presence of accessory cells plays an important role in generating oxidants that attack the target cell. There is also some suggestive evidence that phorbol esters can act directly on B-lymphocytes to generate chromsome breaks (74), but the intriguing question of how B-lymphocytes handle the challenge of exogenous or endogenous oxidants remains to be explored in more detail. For example, when lymphocytes (20) or fibroblasts (82) are cultured with neutrophils and macrophages and then treated with phorbol esters that activate accessory cells, clastogenic effects can be seen in the lymphocytes. In a similar situation where the *in vitro* target cell was 10T 1/2 C3H fibroblasts and the activating system was phorbol ester-stimulated neutrophils, neoplastic transformation was observed as the end point (82). Indomethacin has been found to inhibit the clastogenesis in lymphocytes (20).

The biochemical mechanism of action of indomethacin in accessory cells has not been completely worked out, but this agent has been shown in a number of studies to inhibit the chemiluminescence (radical production) that is produced by neutrophils during phagocytosis (10,31,51,52,79). Oxygen radicals generated by the phagocytic activity of accessory cells could, in turn, attack B-lymphocytes by peroxidation of plasma membrane lipids, thus propagating radical formation to the inside of the B-lymphocyte or plasma cell (Fig. 1). This may create a sufficient load of radicals in B-lymphocytes such that DNA is attacked by an intracellular product of lipid peroxidation.

The inhibitory action of indomethacin in pristane-induced plasmacytomagenesis could then be explained as an inhibition of radical production by the inflammatory phagocytes that are found free in the peritoneal space and/or in the organized oil granuloma. In plasmacytomagenesis the proliferating plasma cells in the foci

---

FIG. 5. Typical plasmacytic focus found in mouse 116 days postpristane. **A:** A focus of hyperchromatic atypical plasma cells is surrounded by oil granuloma (66×). **B:** Mitotic figures can be seen in the higher power magnification (134×). (From ref. 64.)

would be exposed intermittently to potentially mutagenic radicals. This exposure could lead to genetic changes, including chromosomal breaks and base-substitution-type mutations. It is also possible that the clastogenic action of radicals generated in the oil granuloma could be responsible for the nonrandom chromosomal translocations rcpt (12;15), rcpt (6;15), and others. The proliferative plasma cell foci appear to represent an intermediate stage of plasmacytoma development that potentially includes cells with only the first of several biochemical lesions required for the neoplastic progression. If the chromosome 15 translocation, for example, is the primary mutation, then the cells in all foci should have rcpt (12;15) or rcpt (6;15), etc. Studies are in progress on this point.

Growth-Dependent Primary Plasmacytoma Cell

The transition from proliferating plasma cell to a plasmacytoma cell, i.e., one that vigorously overgrows the peritoneal connective tissues and becomes transplantable, is probably prolonged and gradual.

Peritoneal plasmacytomas are diagnosed in the living mouse by finding numerous atypical plasma cells in the peritoneal space. These cells, in doses of 10^5 to 10^6 cells, will not grow progressively if transplanted intraperitoneally into a normal syngeneic mouse (7,59). If, however, the cells are injected into mice that have been conditioned with intraperitoneal pristane, they grow progressively in over 90% of the mice (Table 2). Several primary tumors have been effectively transferred with as few as 10^2 cells (59).

As plasmacytomas are successfully transplanted, they quickly lose the dependence on pristane conditioning, even after a single transfer (39). BALB/c plasma-

TABLE 2. *Effect of pristane conditioning on the intraperitoneal transplantability of primary plasmacytomas*[a]

Inbred strain	Dose of pristane/ml	Treatment	No. of primary tumors	No. of mice with tumors/total	%
BALB/c	0	0	34	7/287	2.4
BALB/c	0.5	0	22	279/283	98
BALB/c	0.5	Cortisone d0	11	32/150	21
BALB/c	0.5	Cortisone d3	11	150/150	100
BALB/c	0.5	Indomethacin d-7	4	19/20	95
BALB/c	0.2	0	4	40/40	100
BALB/c	0.2	Indomethacin d-7	4	40/40	100
BALB/c	0.1	0	4	38/40	95
BALB/c	0.1	Indomethacin d-7	4	34/40	85
BALB/cxDBA/2F$_1$	0.5	0	5	20/25	80
BALB/cxDBA/2F$_2$	0.5	0	2	13/60	21
DBA/2	0.5	0	7	0/35	0

[a]From ref. 7.

cytomas are known to carry relatively strong tumor transplantation antigens (39,50,69,70,83). This characteristic has raised the possibility that the mechanism of action of pristane conditioning is immunosuppressive. Whisson and Conners (83) first showed that the drug aniline mustard could completely destroy large tumor masses (approximately 5 g in size) and render the mice immune to challenge with tumor cells. We subsequently showed that many, but not all, transplanted plasmacytoma cells were similarly able to immunize BALB/c hosts *(M. Potter, unpublished observations)*. In earlier experiments, we also found that pristane conditioning abrogated established immunity to the long-term transplanted plasmacytoma AdjPC5 (60). Mice were first immunized to AdjPC5 by the Whisson-Conners aniline mustard treatment method. After recovery from the toxicity of aniline mustard, the mice were injected intraperitoneally with pristane and then challenged intraperitoneally with live tumor cells. Immune nonpristane-treated mice were better able to reject intraperitoneal challenges of AdjPC5 cells than were immune pristane-treated mice. This result suggested that pristane treatment abrogated cellular immunity. The immunosuppressive action of intraperitoneal injections of mineral oil or pristane has been documented in several other studies (25,36).

Work in our laboratory, however, has suggested that immunosuppression may be only a weak effect and is not the relevant biological action in conditioning the peritoneum. First, pristane is usually injected several days before transplantation, but experiments (M. Potter and J. S. Wax, *unpublished observations)* indicate that pristane administered as long as 10 days after the tumor cells have been injected will still provide the essential conditions for growth. Since delayed conditioning has allowed considerable time for some potential immune processes to eliminate the tumor cells, the results are evidence against the interpretation that pristane conditioning significantly abrogates cytotoxic immune responses. Second, (BALB/c-×DBA/2)F_1 hybrid mice (which are H-2^d/H-2^d) are resistant to plasmacytoma formation initiated by intraperitoneal pristane injections, yet pristane-conditioned F_1 do not reject primary BALB/c plasmacytomas (63). However, 79% of pristane-conditioned (BALB/c×DBA/2)F_2 mice did reject these primary plasmacytomas (63). This partial value suggests that at least three minor histocompatibility gene differences between the BALB/c donor tumor and CDF_2 recipient are sufficient to cause rejection of the primary tumor in the pristane-conditioned mouse. Thus, "minor histocompatibility differences between BALB/c and DBA/2 are not abrogated by pristane conditioning (Table 2). The two findings that suggest pristane does not abrogate immune cytotoxic reactions.

An alternative explanation that we currently favor is that pristane conditioning stimulates the formation of growth factors that promote strong cellular proliferation that overcomes weak immune responses. Evidence that the underlying mechanism of intraperitoneal pristane conditioning is attributable to the production of factors by inflammatory exudate cells was first supported by experiments using continuous hydrocortisone treatment (7). When mice were treated with hydrocortisone prior to intraperitoneal pristane conditioning, many (approximately 77%) were able to reject the primary tumors (Table 2). In contrast, when the hydrocortisone was begun 3 days after pristane conditioning, tumors grew in virtually all of the mice. Hydrocortisone blocked the influx of macrophages into the peritoneum and the formation of the oil granuloma. These experiments raised the possibility that factors supplied by macrophages were required for plasmacytoma growth. Evidence from previous studies by Namba and Hanaoka (47,48) indicated that the *in vitro* growth of the established transplantable plasmacytoma MOPC 104E required the presence of macrophages or factors produced by macrophages for growth. However, the intraperitoneal injection of thioglycollate which stimulates phagocytosis by resident macrophages did not condition the peritoneum for plasmacytoma growth (7).

It has been found that many established transplantable plasmacytomas require macrophage-derived factors for growth *in vitro* and further that many of these tumor cell lines undergo rapid cell death in 24 to 48 hr if deprived of these factors (R. Nordan and M. Potter, *unpublished observations*). Moreover, macrophage-derived factors, called α-factors by Corbel and Melchers (11,12) are required by normal proliferating B-cells (LPS-blasts) for mitosis. Melchers *et al.* (40) have found that cross-linked C3b or C3d can replace the requirement for α-factors in the mitotic cycle. The identity of the plasmacytoma-requiring factors is not known, but they do not appear to be IL1, IL2, IL3, $BCGF_1$ interferon-γ, or TC-GFs (R. Nordan, *unpublished observations*). The relationship of α-factors to the requirement supplied by pristane conditioning is not established. Unlike hydrocortisone, indomethacin does not abrogate conditioning by pristane (Table 2).

ACCELERATORS OF PROGRESSION

Transforming retroviruses can accelerate the progression of plasma cells toward a more aggressive autonomous state. When pristane-conditioned BALB/c mice are infected with Abelson transforming virus (a retrovirus that contains Moloney MuLV and a defective transforming element that carries the v-*abl* oncogene), plasmacytomas can be induced in approximately 10% to 20% of mice with very short latent periods (49). Abelson virus also induces acute pre-B-cell lymphomas in very high frequency in adult BALB/c mice. These occur in more than 80% of the mice, and since they have very short latent periods, many of the mice die of lymphosarcoma. Even though total incidence of plasmacytomas is relatively low, the evidence favoring acceleration is clear (49).

V-*abl* in mice has a tyrosine kinase activity and creates biochemical disorders affecting cellular proliferation (28). The Abelson virus pristane-induced plasmacytomas have v-*abl* proviruses inserted in their genomes and further have nonrandom translocations involving the D2/D3 bands in chromosome 15 (49). This finding suggests that the cells containing the chromosome 15 breaks and rejoinings that deregulate the c-*myc* oncogene are the targets for Abelson virus transformation and that the presence of v-*abl* and the c-*myc* activation in combination transforms the cells.

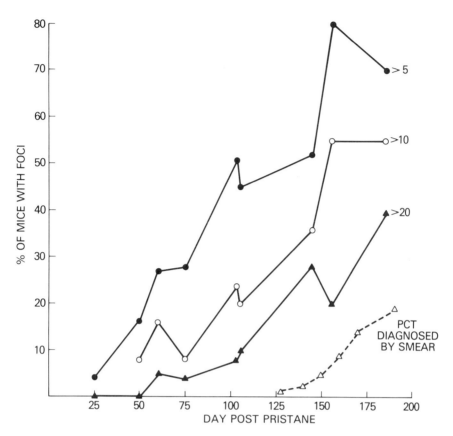

FIG. 6. Development of foci in BALB/c mice after the injection of 0.5 ml pristane at days 0, 60, and 120. The curves depict five or more (>5), 10 or more (>10), or 20 or more (>20) foci. The *dashed line* represents the number of plasmacytomas that were diagnosed by identification of plasmacytoma cells in the peritoneal fluid (62).

Autonomous State

The autonomous state is the transition from conditional to independent growth. Many plasmacytomas show further evidence of instability at this stage, the loss of heavy-chain biosynthesis occurs at an extraordinarily high frequency (42); also, these cells can become non-Ig secretors by the loss of light-chain biosynthesis. Plasmacytoma cells can even undergo downstream productive IgC_H switches (68).

The autonomous plasmacytoma is capable of growing progressively in either intraperitoneal or subcutaneous sites in syngeneic hosts without pristane conditioning. Plasmacytomas, when introduced into a specific site, can infiltrate into connective tissues but rarely metastasize beyond the first chain of lymph nodes. Very slowly growing subcutaneous transplantable plasmacytomas, such as C3H X5563, metastasize to bone marrow cavities if they remain in the host 90 days (56). Only a few BALB/c plasmacytomas have produced osteolytic lesions, chiefly because they grow rapidly and kill the host before metastases are seen. Plasmacytomas that have been extensively transplanted can give rise to more degenerate cell types.

SUMMARY

The peritoneal plasmacytomas of BALB/c mice that are induced by intraperitoneal injections of pristane appear to develop in stages (plateau form of tumor progression). The cell in which the initial neoplastic changes take place has not been identified, but it is probably a mature B-lymphocyte that is in the process of becoming an Ig-secreting cell. The first manifestations of atypical behavior are the foci of proliferating plasma cells that can be found in a few mice as early as 25 days after pristane. These foci continue to appear in more mice and in greater numbers during the latent period (days 0–210). Histological studies strongly suggest that plasmacytomas develop from cells in these proliferative foci. Ninety-eight percent of more than 50 BALB/c plasmacytomas so far examined have chromosomal anomalies (translocations, interstitial deletions) that involve the D2/D3 band region of chromosome 15, the location of the c-*myc* oncogene. These breaks and rejoinings within and near c-*myc* affect transcription and translation by different mechanisms, none of which is yet completely understood. The end result, however, is that c-*myc* is actively transcribed in plasmacytomas. It is generally hypothesized from indirect evidence that c-*myc* gene deregulation affects mitotic cycling. If so, it is tempting to speculate that the deregulation of c-*myc* in plasma cells could constitute a major biochemical lesion that maintains the cells in the cycle, thereby blocking normal maturation to a postmitotic state. Although the time and site of origin of the chromosome 15 translocation has not been determined in plasmacytomagenesis, it is possible tht it is a very early event and may be associated with the proliferative foci. The progression of cells in foci to a stage when they are transplantable to pristane-conditioned hosts appears to require considerable time. Intermediary steps remain to be defined. The mouse plasmacytoma system has the experimental advantage of providing intermediary cell stages for analysis and, when these cells can be isolated and studied, a better understanding of progression may be available.

ACKNOWLEDGMENTS

I thank Ms. Mary Millison and Ms. Victoria Rogers for their patient help in preparing this manuscript. I am grateful to Drs. J. Frederic Mushinski and Emily Shacter for helpful discussions and for critical reading of the manuscript.

REFERENCES

1. Anderson, A.O., Wax, J.S., and Potter, M. (1985): Differences in the peritoneal response to pristane in BALB/cAnPt and BALB/cJ mice. *Current Top. Microbiol. Immunol.*, 122:242–253.
2. Banerjee, M., Wiener, F., Spira, J., Babonits, M., Nilsson, M.-G., Sumegi, J., and Klein, G. (1985): Mapping of the c-*myc*, pvt-1, and immunoglobulin kappa genes in relation to the mouse plasmacytoma variant (6;15) translocation breakpoint. *EMBO J.*, 4:3183–3188.
3. Bazin, H., Beckers, A., Deckers, C., and Moriamem (1973): Transplantable immunoglobulin secreting tumors in rats. V. Monoclonal immunoglobulins secreted by 250 ileocecal immunocytomas in Lou/ws1 rats. *J. Natl. Cancer Inst.*, 51:1359.
4. Bekemeier, H., Bohin, R., Hagen, V., Hannis, E., Henkel, H.-J., Hirschelmann, R., and Wenzel, U. (1982): Structure-activity relationship in non-steroidal anti-inflammatory agents including QSAR in fenamate derivatives. *Agents Actions Suppl.*, 10:17–34.
5. Benner, R., van Oudenaren, A., Bjorklund, M., Ivars, F., and Holmberg, D. (1982): Background immunoglobulin production: measurement of biological significance and regulation. *Immunol. Today*, 3:243–249.
6. Blanchard, J.M., Piechaczyk, M., Dani, C., Chambard, J.C., Franchi, A., Ponyssegur, J., and Jeanteur, P. (1985): C-*myc* gene is transcribed at a high rate in Go arrested fibroblasts and is post-transcriptionally regulated in response to growth factors. *Nature*, 317:443–445.
7. Cancro, M., and Potter, M. (1976): The requirement of an adherent substratum for the growth of developing plasmactyoma cells *in vivo*. *J. Exp. Med.*, 144:1554–1566.
8. Cebra, J.J., Cebra, E.R., Clough, E.R., Fuhrman, J.A., Komisar, J.L., Schweitzer, P.A., and Shahin, R.D. (1983): IGA commitment models for B-cell differentiation and possible roles for T-cells in regulating B-cell development. *Ann. NY Acad. Sci.*, 409:25–37.
9. Cerutti, P.A. (1985): Prooxidant states and tumor promotion. *Science*, 227:375–381.
10. Cheung, K., Archibald, A.C., and Robinson, M.F. (1983): The origin of chemiluminescence produced by neutrophils stimulated by opsonized zymosan. *J. Immunol.*, 130:2324.
11. Corbel, C., and Melchers, F. (1983): Requirement for marcophages or for macrophage or T cell derived factors in the mitogenic stimulation of murine B-lymphocytes by lupopolysaccharides. *Eur. J. Immunol.*, 13:528–533.
12. Corbel, C., and Melchers, F. (1984): The synergism of accessory factors and of soluble α-factors derived from these in the activation of B-cells to proliferation. *Immunol. Rev.*, 78:51–74.
13. Corcoran, L.M., Cory, S., and Adams, J.M. (1985): Transposition of the immunoglobulin heavy chain enhancer to the *myc* oncogene in a murine plasmacytoma. *Cell*, 40:71–79.
14. Cory, S., Gerondakis, S., Adams, J.M. (1983): Interchromosomal recombination of the cellular oncogene c-*myc* with the immunoglobulin heavy chain locus in murine plasmacytomas is a reciprocal exchange. *EMBO J.* 2:697–703.
15. Cory, S., Graham, M., Webb, E., Corcoran, L., and Adams, J.M. (1985): Variant (6;15) translocations in murine plasmacytomas involve a chromosome 15 locus at least 72 kb from the c-*myc* locus. *Embo J.*, 4:675–681.
16. Craig, S.W., and Cebra, J.J. (1971): Peyers patches, an enriched source of precursors for IgA-containing cells in the rabbit. *J. Exp. Med.*, 134:188.
17. Davis, M.M., Kim, S.K., and Hood, L.E. (1981): DNA sequences mediating class switching in α-immunoglobulins. *Science*, 209:1360–1365.
18. Dutton, D.R., and Bowden, G.T. (1985): Indirect induction of a clastogenic effect in epidermal cells by a tumor promoter. *Carcinogenesis*, 6:1279–1284.
19. Dunnick, W., Shell, B.E., and Dery, C. (1983): DNA sequences near the site of reciprocal recombination between a c-*myc* oncogene and an immunoglobulin switch region. *Proc. Natl. Acad. Sci. U.S.A.*, 80:7269–7273.
20. Emerit, I., and Cerutti, P. (1984): Icosanoids and chromosome damage. In: *Icosanoids and Cancer*, edited by H. Thaler-Dao, *et al.*, pp. 127–137. Raven Press, New York.
21. Erikson, J., Miller, D.A., Miller, O.J., Abcarian, P., Skurla, R.M., Mushinski, J.F., and Croce, C.M. (1985): The c-*myc* oncogene is translocated to the involved chromosome 12 in mouse plasmacytoma. *Proc. Natl. Acad. Sci. U.S.A.*, 82:4212–4216.
22. Fahrlander, P.D., Sumegi, J., Yang, J.-Q., Wiener, F., Marcu, K.B., and Klein, G. (1985): Activation of the c-*myc* oncogene by the immunoglobulin gene enhancer after multiple switch region-mediated chromosomal rearrangements in a murine plasmacytoma. *Proc. Natl. Acad. Sci. U.S.A.*, 82:3746–3750.

23. Farber, E. (1984): Precancerous steps in carcinogenesis: their physiological adaptive nature. *Biochim. Biophys. Acta.* 738:171–180.
24. Foulds, L. (1956): The histologic analysis of mammary tumours of mice. II. The histology of responsiveness and progression. The origins of tumours. *J. Natl. Cancer Inst.*, 17:713–753.
25. Freund, Y.R., and Blair, P.B. (1982): Depression of natural killer activity and mitogen responsiveness in mice treated with pristane. *J. Immunol.*, 129:2826–2830.
26. Geldof, A.A., Rijnhart, P., Ende, M., Kors, N., and Langevoort, H.L. (1984) Morphology kinetics and secretory activity of antibody forming cells. *Immunobiology*, 166:296–307.
27. Gerondakis, S., Cory, S., and Adams, J.M. (1984): Translocation of the myc cellular oncogene to the immunoglobulin heavy chain locus in murine plasmacytomas is an imprecise reciprocal exchange. *Cell*, 36:973–982.
28. Goff, S.P., and Baltimore, D. (1982): The cellular oncogene of the Abelson murine leukemia virus genome. *Adv. Viral Oncol.*, 1:127–139.
29. Hann, S.R., Thompson, C.B., and Eisenman, R.N. (1985): C-*myc* oncogene protein synthesis is independent of the cell cycle in human and avian cells. *Nature*, 314:366–369.
30. Hohn, P. (1979): Morphology and morphogenesis of experimentally induced small intestinal tumors. *Curr. Topics Pathol.*, 67:69–144.
31. Horan, T.D., Noujaim, A.A., and McPherson, T.A. (1983): Effect of indomethacin on human neutrophil chemiluminescence and microbicidal activity. *Immunopharmacology*, 6:97–106.
32. Keath, E.J., Kelekar, A., and Cole, M.D. (1984): Transcriptional activation of the translocated c-*myc* oncogene in mouse plasmacytomas: similar RNA levels in tumor and proliferating normal cells. Cell, 37:521–528.
33. Kelly, K., Cochran, B.H., Stiles, C.D., and Leder, P. (1983): Cell specific regulation of the c-*myc* gene by lymphopcyte mitogens and platelet-derived growth factor. *Cell*, 35:603–610.
34. Khan, A.U. (1984): Myeloperoxidase singlet molecular oxygen generation detected by direct infrared electronic emission. *Biochem. Biophys. Res. Commun.*, 122:668–675.
35. Koch, G., Lok, B.D., and Benner, R. (1982): Antibody formation in mouse bone marrow during secondary type responses to various thymus independent antigens. *Immunobiol.*, 163:484–496.
36. Kripke, M.L., and Weiss, D.W. (1971): Studies on the immune responses of BALB/c mice during tumor induction by mineral oil. *Int. J. Cancer*, 422–430.
37. Kunze, E. (1979): Development of urinary bladder cancer in the rat. *Curr. Topics Pathol.*, 67:144–232.
38. Lands, W.E.M., Kulmacz, R.J., and Marshall, P.J. (1984): Liquid peroxidase actions in the regulation of prostaglandin biosynthesis. In: *Free Radical in Biology VI*, pp. 39–61. Academic Press, New York.
39. Mandel, M.A., and DeCosse, J.J. (1972): The effects of heterologous anti-thymocyte sera in mice. V. Enhancement of plasma cell tumor induction. *J. Immunol.*, 109:360–365.
40. Melchers, F., Erdel, A., Schulz, T., and Dierich, M.P. (1985): Growth control of activated synchronized murine B-cells by the C3d fragment of human complement. *Nature*, 317:264–267.
41. Merwin, R.M., and Redmon, L.W. (1963): Induction of plasma cell tumors and sarcomas in mice by diffusion chambers placed in the peritoneal cavity. *J. Natl. Cancer Inst.*, 31:998–1007.
42. Morrison, S.L., and Scharff, M.D. (1981): Mutational events in mouse myeloma cells. *CRC Crit. Rev. Immunol.*, 3:1–22.
43. Morse, H.C.III, Hartley, J.W., and Potter, M. (1980): Genetic considerations in plasmacytomas of BALB/c, NZB, and (BALB/c X NZB)F$_1$ mice. In: *Progress in Myeloma*, edited by M. Potter, pp. 263–279. Elsevier/North Holland, New York.
44. Morse, H.C.III, Pumphrey, J.G., Potter, M., and Asofsky, R. (1976): Murine plasma cells secreting more than the class of immunoglobulin heavy chain. I. Frequency of two or more M-components in ascitic fluids from 788 primary plasmacytomas. *J. Immunol.*, 117:541–547.
45. Morse, H.C.III, Riblet, R., Asofsky, R., and Weigert, M. (1978): Plasmacytomas of the NZB mouse. *J. Immunol.*, 121:1969–1972.
46. Mushinski, J.F., Bauer, S.R., Potter, M., and Reddy, E.P. (1983): Increased expression of *myc*-related oncogene in RNA characterizes most BALB/c plasmacytomas induced by pristane or Abelson murine leukemia virus. *Proc. Natl. Acad. Sci. U.S.A.*, 80:1073.
47. Namba, Y., and Hanaoka, M. (1972): Immunocytology of cultured IgM-forming cells of mouse. I. Requirement of phagocytic cell factor for the growth of IgM-forming tumor cells in tissue culture. *J. Immunol.*, 109:1193–1200.

48. Namba, Y., and Hanaoka, M. (1974): Immunocytology of cultured IgM-forming cells of mouse. II. Purification of phagocytic cell factor and its role in antibody formation. *Cellular Immunol.*, 12:74–84.
49. Ohno, S., Migita, S., Wiener, F., Babonits, M., Klein, G., Mushinski, J.F., and Potter, M. (1984): Chromosomal translocations activating myc sequences and transduction of v-*abl* are critical events in the rapid induction of plasmacytomas by pristane and Abelson virus. *J. Exp. Med.*, 159:1762–1777.
50. Padarathsingh, M.L., Dean, J.H., McCoy, J.L., Lewis, D.D., Northing, J.W., Natori, T., and Law, L.W. (1977): Cell mediated immunity against particulate and solubilized tumor associated antigens of murine plasmacytomas detected by macrophage migration inhibition assays. *Int. J. Cancer*, 20:624–631.
51. Pekoe, G., Van Dyke, K., Mengoli, H., Peden, D., and English, D. (1982): Comparison of the effects of antioxidant non-steroidal anti-inflammatory drugs against myeloperoxidase and hypochlorous and luminol-enhanced chemiluminescence. *Agents Actions*, 12:232–238.
52. Pekoe, G., Van Dyke, K., Peden, D., Mengoli, H., and English, D. (1983): Antioxidation theory of non-steroidal anti-inflammatory drugs based upon the inhibition of luminol-enhanced chemiluminescence from the myeloperoxidase reaction. *Agents Actions*, 12:371–376.
53. Phillips, B. (1980): The genetic effects of oxygen radicals derived from activated phagocytic cells on Chinese hamster fibroblasts. *Life-Chemistry Rep.*, 3:221–228.
54. Piechaczyk, M., Yang, J.Q., Blanchard, J.M., Jeanteru, P., and Marcu, K.B. (1985): Post-transcriptional mechanisms are responsible for accumulation of truncated c-myc RNAs in murine plasma cell tumors. *Cell*, 42:589–597.
55. Pietrangeli, C.E., and Osmond, D.G. (1985): Regulation of B-lymphocyte production in the bone marrow. Role of macrophages and the spleen in mediating responses to exogenous agents. *Cellular Immunol.*, 94:147–158.
56. Potter, M. (1972): Immunoglobulin producing tumors and myeloma proteins of mice. *Physiol. Rev.*, 52:631–719.
57. Potter, M. (1982): Pathogenesis of plasmacytomas in mice. In: *Cancer a Comprehensive Treatise*, edited by F.F. Becker, 2nd ed., Vol. 1, pp. 135–159. Plenum Press, New York.
58. Potter, M. (1984): Genetics of susceptibility to plasmacytoma development in BALB/c mice. *Cancer Surv.*, 3:247–264.
59. Potter, M., Pumphrey, J.G., and Walters, J.L. (1972): Growth of primary plasmacytomas in the mineral oil-conditioned peritoneal environment. *J. Natl. Cancer Inst.*, 49:305–308.
60. Potter, M., Walters, J.L. (1973): Effect of intraperitoneal pristane on established immunity to the Adj-PC5 plasmacytoma. *J. Natl. Cancer Inst.*, 51:875–881.
61. Potter, M., and Wax, J.S. (1981): Genetics of susceptibility to pristane-induced plasmacytomas in BALB/cAn: reduced susceptibility in BALB/cJ with a brief description of pristane-induced arthritis. *J. Immunol.*, 127:1591–1595.
62. Potter, M., and Wax, J.S. (1983): Peritoneal plasmacytomagenesis in mice. A comparison of three pristane dose regimens. *J. Natl. Cancer Inst.*, 71:391–395.
63. Potter, M., and Wax, J.S. (1985): Role of genes in the susceptibility to plasmacytomas. In: *Genetic Control of Host Resistance to Infection and Malignancy (Progress in Leukocyte Biology)*, edited by E. Skameme, Vol. 3, pp. 793–804. Alan R. Liss, Inc., New York.
64. Potter, M., Wax, J.S., Anderson, A.O., and Nordan, R.P. (1985): Inhibition of plasmacytoma development in BALB/c mice by indomethacin. *J. Exp. Med.*, 161:996–1012.
65. Potter, M., Wax, J.S., and Blankenhorn, E. (1985): BALB/c subline differences in susceptibility to plasmacytoma induction. *Current Topics Microbiol. Immunol.*, 122:234–241.
66. Potter, M., Wiener, F., and Mushinski, J.F. (1984): Recent developments in plasmacytomagenesis in mice. *Advan. Viral Oncol.*, 4:139–162.
67. Rabbitts, P.H., Watson, J.V., Lamond, A., Forster, A., Stinson, M.A., Evan, G., Fischere, W., Atherton, E., Sheppard, R., and Rabbits, T.H. (1985): Metabolism of c-*myc* gene products: c-*myc* mRNA and protein expression in the cell cycle. *EMBO J.*, 4:2009–2015.
68. Radbruch, A., Liesegang, B., and Rajewsky, K. (1980): Isolation of variants of mouse myeloma X63 that express changed immunoglobulin class. *Proc. Natl. Acad. Sci. U.S.A.*, 77:2909–2913.
69. Rohrer, J.W., Vasa, K., and Lynch, R.G. (1977) Myeloma cell immunoglobulin expression during *in vivo* growth in diffusion chambers. Evidence for repetitive cycles of differnetiation. *J. Immunol.*, 119:861.

70. Rollinghoff, M., Rouse, B.T., and Warner, N.L. (1973): Tumor immunity to murine plasma cell tumors. I. Tumor associated transplantation antigens of NZB and BALB/c plasma cell tumors. *J. Natl. Cancer Inst.*, 50:159–172.
71. Roux, M.E., McWilliams, M., Phillips-Quaghata, J.M., and Lamm, M.E. (1981): Differentiation pathway of Peyers patch precursors of IgA plasma cells in the secretory immune system. *Cellular Immunol.*, 61:141–153.
72. Sainte-Marie, G. (1966): Cytokinetics of antibody formation. *J. Cell. Physiol.*, 67 (Suppl. 1):109–128.
73. Scherer, E. (1984): Neoplastic progression in experimental hepatocarcinogenesis. *Biochim. Biophys. Acta.*, 738:219–236.
74. Shiraishi, Y. (1985): Bloom syndrome B-lymphoblastoid cells are hypersensitive towards carcinogen or tumor promoter induced chromosomal alterations and growth in agar. *Embo. J.*, 4:2553–2560.
75. Smeland, E., Godal, T., Rund, E., Beiske, K., Funderud, S., Clark, E.A., Pfeiffer-Ohlsson, S., and Ohlsson, R. (1985): The specific induction of myc protooncogene expression in normal human B-cells is not a sufficient event for acquisition of competence to proliferate. *Proc. Natl. Acad. Sci. U.S.A.*, 82:6255–6259.
76. Thompson, C.B., Challoner, P.B., Neiman, P.E., and Groudine, M. (1985): Levels of c-*myc* oncogene mRNA are invariant throughout the cell cycle. *Nature*, 314:363–369.
77. Tseng, J. (1984): A population of resting IgM-IgD double bearing lymphocytes in Peyers patches: the major precursor cells for IgA plasma cells in the gut lamina propice. *J. Immunol.*, 132:2730–2735.
78. Van Ness, B.G., Shapiro, M., Kelley, D.E., Perry, R.P., Weigert, M., D'Eustachio, D., and Ruddle, F. (1983): Aberrant rearrangement of the κ light chain locus involving the heavy chain locus and chromosome 15 in a mouse plasmacytoma. *Nature*, 301:425–427.
79. Van Dyke, K., Van Dyke, C., Udeinya, J., Brister, C., and Wilson, M. (1979): A new screening system for nonsteroidal anti-inflammatory drugs based on inhibition of chemiluminescence produced from human cells (granulocytes). *Clin. Chem.*, 25:1655–1661.
80. Warner, N.L. (1975): Review. Autoimmunty and the pathogenesis of plasma cell tumor induction in NZB and hybrid mice. *Immunogenetics*, 2:1–20.
81. Webb, E., Adams, J.M., and Cory. S. (1984): Variant (6;15) translocation in a murine plasmacytoma occurs near an immunoglobulin κ gene but far from the myc oncogene. *Nature*, 312:777–779.
82. Weitzman, S.A., Weitberg, A.B., Clark, E.P., and Stossel, T.P. (1985): Phagocytes as carcinogens: malignant transformation produced by human neutrophils. *Science*, 227:1231–1237.
83. Whisson, M.E., and Connors, T.A. (1965): Drug induced regression of large plasma cell tumours. *Nature*, 205:406.
84. Wiener, F., Ohno, S., Babonits, M., Sumegi, J., Wirschubsky, Z., Klein, G., Mushinski, J.F., and Potter, M. (1984): Hemizygous interstitial deletion of chromosome 15 (band D) in three translocation negative murine plasmacytomas. *Proc. Natl. Acad. Sci. U.S.A.*, 81:1159–1163.
85. Yang, J.-Q., Bauer, S.R., Mushinski, J.F., and Marcu, K.B. (1985): Chromosome translocations clustered 5' of the murine c-myc gene qualitatively affect promoter usage. Implications for the site of normal c-myc regulation. *EMBO J.*, 4:1441–1447.
86. Yunis, J.J., and Soreng, A.L. (1984): Constitutive fragile sites and cancer. *Science*, 226:1199–1204.

Multistep Model of Mouse Mammary Tumor Development

David W. Morris and Robert D. Cardiff

Department of Pathology, University of California, Davis, California 95616

Tumor biologists generally agree that neoplastic development (32) is a progressive process involving multiple "steps." Progression is presumably mediated by a complex network of interacting genes (43) as well as other epigenetic and developmental factors (66). The end-stage tumor is the product of consecutive as well as collateral changes. Although considerable progress has been made toward cataloging these changes (7,76), relatively little is known of their temporal or causal relationships. These relationships can be explored with model systems, such as the mouse mammary tumor system, in which sequential morphological stages of neoplastic development can be analyzed.

In this chapter, we discuss molecular studies of multistage mouse mammary tumor development. A general review of the molecular biology of mouse mammary hyperplasia is also available (11).

HYPERPLASIAS AND TUMOR DEVELOPMENT

Several distinct mammary hyperplasias are precursors of adenocarcinomas in the mouse (11,14,48,51). Hyperplastic alveolar nodules (HANs) and plaques are the best characterized. These intermediate-stage lesions represent different morphological pathways of tumor development rather than different morphological stages from a single sequential process. Transformation is proposed to occur in at least two steps in the HAN pathway: nodule transformation of normal cells to HANs, and tumor transformation of nodule cells to tumors (21). We refer to the entire process as *neoplastic progression*. In this context, *tumor progression* refers to changes subsequent to tumor transformation.

HANs are hyperplasias of the lobuloalveolar epithelium (35). They are morphologically and histologically similar to the lobuloalveolar units of prelactating mammary glands. In contrast to normal lobuloaveolar epithelium, HANs are maintained by constitutive levels of lactogenic hormones (6) and appear as local hyperplasias in glands from nulliparous or nonpregnant, nonlactating multiparous females (Fig.

1). Several lines of evidence indicate that HANs are mammary tumor precursors. The most direct evidence is the histological observation of hyperplasia and carcinoma *in situ* within the same lesion. In addition, HANs are induced by the same agents that induce mammary tumors (51), i.e., mouse mammary tumor virus (MMTV), hormones, and chemical carcinogens. A strong correlation also exists between the presence of HANs in the mammary gland and eventual tumor development (39).

Transplantation experiments reinforce the idea that HANs are precursor lesions. Transplants of HANs into the gland-free mammary fat pads of syngeneic mice (see below) produce tumors at a higher frequency and with a shorter latency period than similar transplants of normal mammary epithelium (20). This is convincing biological evidence that transplanted HANs are tumor precursors. However, removal of a HAN from its normal microenviornment releases the hyperplasia cells from the regulatory control of the surrounding normal parenchyma (30). Therefore, whether or not transplanted HANs are biologically equivalent to HANs *in*

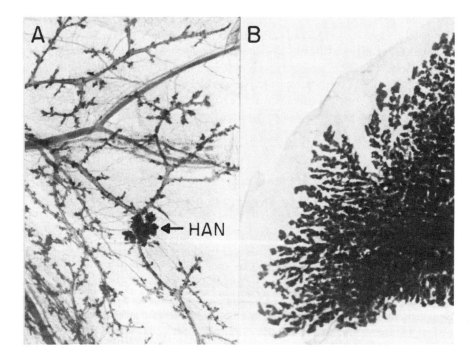

FIG. 1. Morphological comparison of normal and hyperplastic mammary tissues. **A:** Normal unstimulated mammary gland. The unstimulated mammary gland is a branching ductal system. When stimulated with lactoogenic hormones, as in the prelactating or lactating mammary gland, extensive lateral lobuloalveolar development occurs. This growth normally regresses after lactation. A HAN is seen in the unstimulated mammary gland as a focal lesion of lobuloalveolar development maintained by constitutive levels of hormones. **B:** HOG. When HAN tissue is transplanted into the gland-cleared fat pad of an unstimulated host, it proliferates to form a HOG morphologically similar to stimulated prelactating mammary gland. 10× magnification.

situ is still open to question. In addition, selective pressures during outgrowth of transplanted HANs could also confer additional tumor potential to the tissue.

Plaques are disc-shaped, organoid hyperplasias. They are approximately 10 times the size of a HAN at first observation (31). These lesions are remarkable because they arise during pregnancy, regress at parturition, and reappear at the next pregnancy. Plaques are usually classified as small, hormone-dependent tumors. They are referred to as hyperplasias in this chapter to emphasize their biological similarities to HANs. Both HANs and plaques are intermediate-stage, group B lesions by the nomenclature of Foulds (32).

Other less well-studied abnormal hyperplasias and dysplasias also occur in the mouse mammary gland (22,31). Some are end-point lesions with no obvious neoplastic potential, whereas others are proposed tumor precursors. Chemically induced ductal hyperplasias are tumor precursors by both histological and transplantation criteria (50). Keratinized nodules have been proposed as precursors of MMTV-free, urethan-induced adenoacanthomas in the DD strain (41).

TRANSPLANTATION OF HYPERPLASIAS

Biochemical analysis of intermediate stages of tumor development became practical with the development of a system for serial transplantation of HANs (20). The system is particularly valuable in light of the difficulties of growing mouse mammary cells in long-term cell culture (24) and distinguishing among normal, hyperplastic, and tumor cells in short-term primary cultures (78). HAN transplantation is, in effect, a system for "*in vivo* cell culture." Details of the transplantation system are found in DeOme et al. (20) and in a review by Medina (48). HAN tissue is transplanted into the inguinal mammary fat pads of 3-week-old, virgin hosts. The host mammary gland anlage, which only fills 10% to 20% of the fat pad at 3 weeks of age, is surgically removed prior to transplantation. The transplanted tissue proliferates to fill the gland-cleared mammary fat pad and is called a hyperplastic outgrowth or HOG (Fig. 1). Host tissues are clearly distinguishable from transplant tissues: The mammary parenchyma is transplant-derived, and the mesenchymal elements are host-derived (48). HAN transplants grow to form lobuloalveolar structures similar to those in prelactating and exogenous hormone-stimulated mammary glands.

Normal mammary glands, plaques, and various dysplasias and other hyperplasias can also be transplanted. Each tissue behaves differently. Normal mammary gland transplants proliferate to form ductal structures similar to those in unstimulated mammary glands (19). Plaque transplants form dysplastic ductal structures. If animals bearing plaque transplants are stimulated with exogenous hormones, hormone-dependent tumors appear (1).

Comparison of normal mammary outgrowths and HOGs reveals an important distinction. The growth rate of normal outgrowths declines during serial transplantation, and the lines are lost by generation 7 (19). HOGs, on the other hand, are immortal (19). HOG tissues can be serially transplanted to establish HOG lines, such as the C (49), D (52), and Z (3) series of outgrowths from HANs induced by DMBA, hormonal stimulation, and MMTV, respectively. Several lines have been in transplant for more than 70 generations. Although HOGs are immortal tissues, they retain a relatively normal organizational structure and are still under local growth regulation. HOGs transplanted into gland-free fat pads grow to fill the host pads, then cease growth and do not invade surrounding tissues or metastasize (30). If the host fat pads are not cleared of parenchyma prior to transplantation, HOGs will grow to fill the available fat pad but will not overgrow host ductal tissue (30). HOGs transplanted outside the fat pad are maintained but do not grow. These properties of HOGs are in contrast to those of transplanted mammary tumors, which will grow outside the fat pad, will overgrow host mammary tissues, and are locally invasive and metastatic.

Tumors appear in HOGs more frequently than in outgrowths of normal mammary glands (20). A critical point is that a HOG-derived tumor is first observed as a small focal mass in a field of outgrowth. Such tumors can be easily dissected from the surrounding hyperplasia, providing samples from sequential stages of the development of an individual neoplasm. The ability to study progression within an individual lineage is important because of the considerable intersample variability displayed by mammary neoplasms.

EXOGENOUS MMTV PROVIRAL DNA AS A MARKER FOR CLONAL SUBPOPULATIONS DURING NEOPLASTIC PROGRESSION

Hyperplasias originate as focal lesions in normal mammary glands. Likewise, tumors originate as focal lesions in hyperplasias. This suggests a process of neoplastic progression by selection of subclones with enhanced growth potential. Molecular proof of clonal selection and multistage neoplastic progression (Fig. 2) was provided in this system by using MMTV as a marker for clonal cell populations.

MMTV is the major etiologic agent in sporadic mouse mammary cancer (5,55). The virus is transmitted horizontally through the milk (8,70,77) or germinally via active endogenous proviruses (4). As part of the lifecycle of the retrovirus, the RNA genome is reverse transcribed, and a proviral DNA copy is inserted into the mouse genome. Integration of proviral DNA is presumably random. Each infected cell bearing exogenous proviral DNA has a unique "fingerprint" from the positions of its newly acquired proviruses. In practical terms, a clonal population of cells bearing exogenous proviruses has restriction fragments, unique in size, which span the host genome/virus genome junctions. The unique fragments are detectable by the Southern blot technique using radiolabeled MMTV probes (15). For mixed populations of cells, a restriction fragment must be present in at least 5% of the total cell population to be readily detected by the Southern blot technique. Comparisons of the intensities of bands from endogenous proviruses with the intensities of bands from novel junction fragments allow an estimate of the fraction of the total cell population carrying a given exogenous provirus. By these criteria of Southern blot analysis and detectable novel MMTV restriction fragments, all tu-

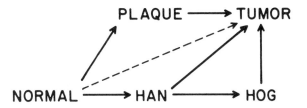

FIG. 2. Pathways of morphological progression during mouse mammary tumorigenesis. HANs are abnormal lesions that can be serially transplanted indefinitely in gland-free mammary fat pads as HOGs. Tumors arise as focal lesions in the outgrowths. Other lesions, such as ductal hyperplasias (not shown), are also tumor precursors that can be maintained as outgrowths. Plaques are organoid, tumor-like lesions proposed to arise independently of HANs in some strains. It is not known if tumors can arise without progressing through a hyperplastic stage *(dashed arrow)*.

mors from virus-positive strains contain exogenous proviruses and clonal cell populations (26).

HANs are Clonal

HANs and HOGs have been similarly analyzed by Southern blot analysis to determine whether HANs are nonclonal hyperplasias or clonal proliferations similar to the tumors toward which they evolve (2,12,13,28,68). One HAN has been successfully studied and had detectable unique host/virus junction fragments (13), suggesting that HANs are clonal proliferations. Considerable support for this idea has come from analyses of HOGs (12,13), which are larger and therefore easier to study. Like HANs, HOGs also have clonal exogenous proviruses. In addition, HOGs from a single, subdivided HAN always share novel host/virus junction fragments. This demonstrates that at least a fraction of the cells within each related subline, and hence within the progenitor HAN, have a common, clonal origin. It cannot be determined whether HANs are monoclonal or whether they originate from multiple cells with a few clonal lineages growing to dominate the hyperplastic cell population. Therefore, HANs are clonal but not necessarily monoclonal.

In light of the biological characteristics of HANs (i.e., genetically altered, clonal, and immortal), and because they are the earliest morphologically or biochemically identifiable lesions in the continuum of neoplastic progression in the mammary gland, we have proposed the term "protoneoplastic," rather than "preneoplastic," to characterize the HAN (11).

Neoplastic Progression

Similar studies of clonality using samples from sequential morphological stages of tumor development have demonstrated neoplastic progression at a molecular level (13). As indicated above, HOGs and focal tumors dissected from HOGs can be directly compared. Southern blot analyses of paired HOG and HOG-derived tumor samples demonstrate that outgrowth cells progress to tumor cells (Fig. 3). The MMTV fingerprint of the HOG is entirely retained by the tumor, in most cases. Also, additional tumor-specific clonal MMTV restriction fragments usually appear (Fig. 3, panel 1) although exceptions are observed (Fig. 3, panel 3).

Growth selection of genotypic variants with increased autonomy is the proposed mechanism of neoplastic progression. Variant lineages are often characterized by divergent MMTV restriction patterns. Most data indicate that the divergent patterns reflect an increase in provirus copy number due to superinfection or transpo-

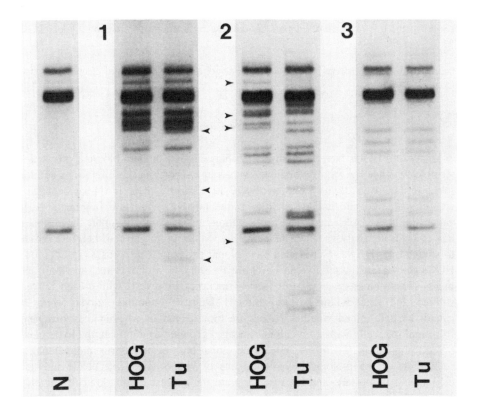

FIG. 3. Exogenous MMTV proviral restriction patterns as markers for clonal cell populations during neoplastic progression. Shown are Southern blot analyses of paired DNA samples from HOGs and HOG-derived tumors digested with Bam HI and probed with radiolabeled MMTV LTR. Three classes of restriction pattern evolution are observed when HOG DNA is compared with DNA from a discrete focal tumor dissected from the outgrowth: *(1)* conservation of the progenitor HOG pattern and acquisition of additional proviruses in the tumor; *(2)* emergence of a tumor from a subpopulation in the HOG with resultant loss of a portion of the HOG restriction pattern (type 1 changes are also seen); *(3)* no perturbations of MMTV restriction patterns (verified with other enzymes), suggesting tumor progression in the absence of additional acquisition of exogenous provirus. All HOG samples are primary outgrowths from BALB/cfC3H HANs. N is normal control DNA.

sition in cells of the lineages that ultimately grow to dominate the cell population. Changes in lineages that grow to make up less than approximately 5% of the final population are undetectable. Some exogenous proviruses are nonrandomly integrated in tumors and may be factors in this growth selection (58,61; D.W. Morris and R.D. Cardiff, *unpublished data*). Other exogenous proviruses are probably not involved in selection but are observed by Southern blot analysis as a consequence of their chance integration in a cell under selection by other factors. Deletions or rearrangements of integrated MMTV proviruses have not been reported, suggesting that genomic instability of integrated proviruses is not a factor in most of the observed changes.

A clear example of clonal selection is shown in Fig. 3, panel 2. A direct comparison of a primary HOG and a focal tumor removed from the HOG reveal that some of the exogenous restriciton fragments present in the HOG are retained in the ensuing tumor, some are lost, and several restriction fragments are present only in the tumor. This is evidence that multiple divergent clonal subpopulations (with divergent MMTV restriction patterns) having a common origin were present within the primary HOG population (13). A cell from one of the subpopulations progressed to a tumor with a concomitant appearance of new clonally integrated, exogenous proviruses.

Additional evidence of clonal selection during neoplastic progression was obtained in studies of tumor progression from hormone dependence to hormone independence in the GR strain (73). Hormone-dependent tumors require estrone and progesterone to grow in castrated mice (69). The tumors progress during serial transplantation to partially autonomous tumors that grow in the absence of exogenous hormones but at a reduced rate. Ultimately, the tumors become completely independent of hormones for growth. Using exogenous MMTV restriction patterns as markers for cell populations in this system, MacInnes et al. (47) and Michalides et al. (53) were able to identify clonal subpopulations in hormone-dependent GR tumors and observed clonal selection of subpopulations during progression of these tumors to hormone independence during serial passage. Gradual changes in the intensities of various MMTV restriction fragments were seen in biopsy samples from each successive cycle of growth and regression. These changes were presumably correlated with the loss of hormone-dependent cells and the appearance of more autonomous cell populations.

TUMOR-ASSOCIATED MMTV INTEGRATION LOCI

Attempts to identify genes involved in mammary cancer have focused on the chromosomal sites of MMTV integration. The hypothesis is that integration of exogenous MMTV proviral DNA into the host genome can, in rare cases, activate expression of a cellular mammary oncogene: the so-called insertion mutagenesis model developed in the avian leukosis virus system (37,75). Efforts to clone this type of gene have thus far yielded two candidates.

Cellular Genes Activated by MMTV

The experimental strategy used to find genes activated by MMTV is to identify a tumor sample with a single newly integrated exogenous provirus, clone host DNA that flanks the provirus, and identify the cellular gene activated by provirus insertion. The tacit assumptions of this approach are that (a) integration of exogenous MMTV proviruses is random, or effectively random, throughout the mouse genome; (b) rare provirus integrations can activate expression of cellular genes involved in mammary growth control; and (c) each MMTV-induced tumor has undergone at least one insertion activation event. The model accounts for the long latency period of MMTV–induced tumors, the finding that many mammary cells are infected with MMTV but few become transformed, and the clonal nature of the newly acquired exogenous proviruses in MMTV-induced tumors (13,15,16,25,27,34). This strategy has led to the identification of two different loci harboring putative oncogenes: *int-1* (58) on chromosome 15 (57,62) and *int-2* (61) on chromosome 7 (62).

The *int-1* gene was identified by Nusse and Varmus (58) within a tumor-associated MMTV integration locus approximately 20 kilobases (kb) in length. Exogenous proviruses were found integrated at this locus in 18 of 26 MMTV-induced tumors from the C3H strain (58). A 2.6-kb messenger RNA from the *int-1* gene was detected in tumors with *int*–provirus (57,58). By hybridization (58) and DNA sequence (74) comparisons, the *int-1* gene has no detected homology to known oncogenes.

The *int-2* gene was identified by Peters et al. (61) within a different tumor-associated MMTV integration locus similar in size to the region harboring *int-1*. Exogenous proviruses were detected in the *int-2* region in 21 of 39 MMTV-induced tumors from the BR6 strain. These results were confirmed by an independent isolation of the *int-2* locus in our laboratory (D.W. Morris, H.D. Bradshaw, Jr., and R.D. Cardiff, *unpublished data*). The predominant messenger RNA from the *int-2* gene is 3.2 kb long, with several unexplained minor transcripts (23). As with the *int-1* gene, transcription of *int-2* has been reported in tumors bearing a provirus in the *int-2* region, with no detectable messenger RNA in normal tissues and most *int-2* provirus-free tumors. *Int-2* also has no detectable homology to known oncogenes (61).

Several results suggest that these genes are legitimate mammary oncogenes and therefore that this technique of identifying genes is a valid means for determining some of the multiple factors that must contribute to multistep tumor development. First, no proviruses integrated in the coding regions of the *int-1* gene have been observed. Those proviruses integrated in the noncoding regions of the first and last exons are integrated so that proviral regulatory sequences replace the *int-1* promotor or termination sequences, respectively (74). Thus, the functional integrity of the gene is always maintained. Similarly, a portion of the *int-2* region, known to be transcribed, is always free of provirus (23). Second, no tumors have been observed that have both homologs of the *int-1* or *int-2* loci disrupted by provirus integration. The total number of clonally integrated exogenous proviruses in the samples surveyed is large relative to the number of proviruses observed at *int-1* or *int-2*. Statistically, if these loci were "targeted" for MMTV integration in some specific fashion, tumors with integrations in both homologs should be observed. None has been reported, supporting the assumption that observation of proviruses at *int-1* or *int-2* in tumors is a consequence of clonal selection due, in part, to enhanced *int-1* or *int-2* expression. Finally, and perhaps most compelling, is the paradigm of ALV activation of the c-*myc* gene (36,56,59,60). ALV, the "prototypic virus" of the slow-transforming retrovirus class, activates a known cellular oncogene, c-*myc*, by insertion mutagenesis in a manner similar to MMTV activation of *int-1* or *int-2*. Both *int-1* and *int-2* genes can be activated (a) in *cis* (23,57,74) presumably by the enhancers in the MMTV long terminal repeats (LTRs) (64), when proviruses integrate in the chromosomal regions flanking the genes, or (b) by promoter insertion (23,74), when MMTV integrates in the noncoding portion of the first exon so that the 3' LTR promoter substitutes for the natural promoter. The MMTV and ALV systems vary in the observed frequencies of promoter insertion versus "enhancer insertion." The latter is predominantly observed in MMTV-induced tumors.

Int-1 and *Int-2* in Neoplastic Progression

Int-1 and *int-2* are not dominant transforming genes. Integration of provirus at either locus is not sufficient for malignant transformation. Peters et al. (63) observed *int-2* integrations in two of three hormone-dependent BR6 tumors. Because these tumors fully or partially regress between pregnancies, it was concluded that at least one additional event was required for frank malignancy. In our laboratory, five of nine plaques surveyed had a provirus at the *int-2* locus (Table 1). In addition, a BALB/cfC3H HOG line has been developed that bears a MMTV provirus integrated at the *int-1* locus (Fig. 4B). The provirus was detected in only one of two HOG sublines derived from a primary HOG and probably appeared by clonal selection during outgrowth of the second transplant generation. The *int-1* integration had no apparent effect on the morphology of the HOG line. These results demonstrate that activation of *int-1* or *int-2* is not sufficient for malignant transformation. The genes must act in concert with other factors as predicted by the multistep model.

Samples from sequential stages of neoplastic progression have given additional hints of the role of these genes in tumorigenesis. A survey of tumors and HOGs for *int-1* and *int-2* activation from our laboratory is summarized in Table 1. Two conclusions can be drawn. First, each tissue type has a characteristic probability of *int-1* or *int-2* activation. In some classes of MMTV-induced tumors and hyperplasias, these two genes appear to play little or no role. The implication is that other important MMTV-activated genes remain to be discovered. Second, *int-1* and *int-2* activation occurs during tumor transformation or progression in the HOG transplantation system (Fig. 4). The failure to detect activation of *int-1* or *int-2*

TABLE 1. *Survey of MMTV-induced mammary hyperplasias and tumors for provirus integrations at the int-1 and int-2 loci*[a]

Strain	Tissues	int-1 Integrations[c]	int-2 Integrations[c]
GR	HOGs	0/8	0/8
	HOG-derived tumors	4/24	2/24
	Plaques	0/9	5/9
	Spontaneous tumors	7/27	12/27
BALB/c-GR	HOGs	0/4	0/4
	Spontaneous tumors	3/9	1/9
C3H	HOGs	0/6	0/6
	HOG-derived tumors	1/14	1/14
	Spontaneous tumors	9/9	1/9
BALB/cfC3H	HOGs	1[b]/7	0/7
	HOG-derived tumors	5/23	1/23
	Spontaneous tumors	7/18	3/18

[a]D. W. Morris, R. Strange, P. A. Barry, V. K. Pathak, L. J. T. Young, H. D. Bradshaw, Jr., and R. D. Cardiff, *unpublished data*.

[b]Appeared during transplantation (see text).

[c]As determined by Southern blot analyses for *int-1* and *int-2* restriction fragments interrupted by MMTV integration using the following probes [*int-1* (58); *int-2* (unpublished data)] and enzymes:

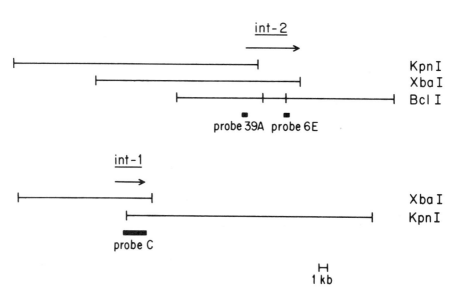

during nodule transformation supports a model in which *int-1* and *int-2* function exclusively, at least in the transplantation system, subsequent to nodule transformation (immortalization).

Considerably more information must be gathered before a function in neoplastic progression can be assigned to the *int-1* and *int-2* genes. We envision a set of *int*

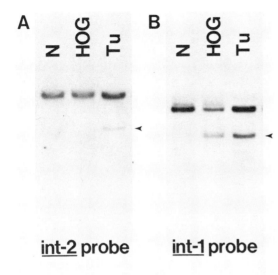

FIG. 4. Activation of putative mammary oncogenes during neoplastic development. A ninth generation HOG from a BALB/cfC3H HOG line and a tumor from the HOG line were compared by Southern blot analysis. DNA samples were digested with Xba I and sequentially probed with (A) int-2 probe 39A and (panel B) int-1 probe C (58). A clonal integration of MMTV provirus at the int-1 locus is present in the HOG and maintained in the tumor (B). Disruption of the int-2 locus by integration of MMTV provirus is seen in a subpopulation of the tumor cell population (panel A), perhaps occurring as a post-transformation event during tumor progression. (R. Strange, *unpublished data.*)

genes with a spectrum of "neoplastic potentials" reflecting, for example, each gene's position in a regulatory cascade or the scope of its pleiotropic effects on other growth-related genes. Currently, however, the number of genes of this type that may be involved in mammary tumorigenesis is not clear. New host-flanking sequences cloned from samples with proviruses proposed to be integrated in tumor-associated MMTV integration loci are currently being characterized in several laboratories.

OTHER GENES ASSOCIATED WITH MAMMARY CANCER

Numerous other genes associated with mammary cancer have been identified in the past 5 years in human, rat, and mouse samples. To the extent that the human- and rat-associated genes may also prove important in mouse mammary cancer, the genes implicated in these systems, as well as mouse, are listed in Table 2. The list will undoubtedly grow in the next several years. These known and putative mammary oncogenes have been identified by three experimental approaches: (a) isolation of genes active in various DNA-mediated gene transfer assays for dominant transforming mutations, (b) construction of transgenic mice bearing oncogenes, and (c) direct comparisons of normal and tumor tissues and cell lines.

Transformation Assays

The prevailing model of chemical carcinogenesis proposes that dominant somatic mutations of cellular oncogenes are responsible for transformation (17,79). The NIH 3T3 transformation assay (67) was designed to detect such mutations. Several innovations (outlined below) have broadened the utility of the assay, allowing study of epithelial transformation and detection of novel oncogenes.

TABLE 2. Cellular oncogenes and putative oncogenes associated with human, rat, and mouse mammary cancers

Gene	Species	Source	Mode of activation	Ref.
c-erbB-2	Human	Primary tumor	Amplification	42
int-1	Mouse	MMTV-induced primary tumors	MMTV-insertion mutagenesis	58
int-2	Mouse	MMTV-induced primary tumors	MMTV-insertion mutagenesis	61
mcf2	Human	MCF-7 cell line	Unknown[a]	29
mcf3	Human	MCF-7 cell line	Unknown[a]	29
c-myc	Human	SKBR-3 cell line	Amplification	44
	Human	SW 613-S cell line	Amplification	54
MTV/myc	Mouse (transgenic)	Primary tumors	Heterologous promoter	71
c-Ha-ras1	Human	HS578T carcinosarcoma cell line	Spontaneous codon 12 mutation	45
	Rat	NMU-induced tumors	Codon 12 mutation	72,80
	Rat	DMBA-induced tumors	Codon 61 mutation	80
	Mouse	DMBA-induced tumors from a hormone-induced HOG line	Codon 61 mutation	18
N-ras	Human	MCF-7 cell line	Amplification	29,33
tx-1	Human	MCF-7 cell line	Unknown[a]	46
	Mouse	DMBA-induced tumors from a hormone-induced HOG line and MMTV-induced tumors	Unknown[a]	46

[a] Identified using the NIH 3T3 transfection assay for dominant transforming genes.

Transforming genes were detected (18) using the NIH 3T3 assay in four of four mammary tumors from DMBA treated animals bearing the UCD variant (2) of the hormone-induced D1 hyperplastic outgrowth line (52). All samples had mutations in the second position of codon 61 of c-Ha-*ras*1 (18). DNAs from spontaneous tumors arising in untreated D1 hyperplastic outgrowths did not transform NIH 3T3 cells. In addition, another transforming gene provisionally named tx-1 (76), was also detected in human MCF-7 cells and virus and DMBA-induced mouse mammary tumors by the NIH 3T3 assay (46). The gene has not been isolated.

The NIH 3T3 assay is useful for identifying some transforming genes but is not an adequate model for epithelial transformation. An objection is that the assay seems to mainly detect mutated *ras* genes. This raises the possibilities that the assay is (a) insufficiently sensitive to detect weakly transforming oncogenes and (b) unable to detect tissue-specific oncogenes from various nonfibroblastic cell types. In response to these concerns, a number of investigators have adopted tumor formation in nude mice as a new transformation criteria for transfected cells (10,29) and/or have switched to mammary cells as DNA recipients (40).

Detection of mammary cell-specific oncogenes by using mouse mammary cells as DNA recipients has not yet been reported. However, a study demonstrating that c-Ha-*ras*1 is mitogenic in a "normal" mouse mammary cell line (40) supports the results from the NIH 3T3 assay of DMBA-induced tumors. Nontumorigenic NMuMG cells derived from a normal NAMRU mouse mammary gland became tumorigenic in nude mice following transfection with activated c-Ha-*ras*1 DNA cloned from the human EJ bladder carcinoma cell line.

Transgenic Mice

Development of transgenic mice bearing oncogenes linked to tissue-specific enhancers is a new experimental system that potentially will allow reconstruction of postulated transformation mechanisms *in vivo*. The tissue-specific enhancer of MMTV (64,65) makes such experiments in the mouse mammary tumor system possible. Stewart et al. (71) generated 13 transgenic mouse lines, each bearing an MMTV LTR/*myc* fusion gene. In two of the lines in which the c-*myc* promoters were replaced by the MMTV LTR promoter and glucocorticoid hormone receptor binding site, mice developed spontaneous adenocarcinomas during early pregnancies. The fusion gene alone, however, is not sufficient for tumorigenesis. The reported tumors appeared only during pregnancy and as single, focal lesions in each mouse with most of the mammary glands unaffected.

DISCUSSION AND SUMMARY

Tumors are heterogeneous, evolving tissues with properties that may not accurately reflect those of the incipient neoplasms. The use of tumors to study transformation presupposes that the key transforming "events" are conserved and recognizable amid a host of ancillary changes arising during tumor progression (38). The problem with this approach is distinguishing cause from effect. Our current understanding of neoplastic progression can be enhanced by the use of lesions that represent intermediate stages of tumor development.

Several hyperplastic lesions have been identified that are intermediate stages in mammary tumorigenesis. Biochemical analysis of these hyperplasias became possible with the development of the hyperplasia transplantation system. The major asset of this system is that it provides a means to categorize molecular events during mammary neoplastic progression into potential initiation events (occurring during nodule transformation) and progression events (occurring during, or subsequent to, tumor transformation). Because neoplasms develop in multiple steps, the ability to establish temporal relationships among phenomena is important in determining the mechanisms of transformation.

The induction of tumors and hyperplasias by MMTV has made the molecular analysis of the multistage mouse mammary tumor system possible Exogenous MMTV can be used as a marker to identify and track clonal cell populations during neoplastic progression. More important, exogenous MMTV, acting as an insertional mutagen, has pointed to some of the host genes that may drive the process of tumor development. Alternate molecular approaches have identified even more potential mammary oncogenes.

In this review, we have discussed the initial molecular characterization of the mouse mammary tumor system. Although our understanding of mammary tumorigenesis is incomplete, several generalizations can be made:

1. Mouse mammary tumorigenesis is a multistep process. Several intermediate stages have been identified by morphological, histological, and hormonal criteria. Molecular studies demonstrate that some of these lesions are intermediate stages in neoplastic progression.

2. Mammary hyperplasias, in general, are clonal and contain divergent subpopulations of genotypic variants.

3. Clonal selection of genotypic variants occurs throughout mammary neoplastic progression.

4. Activation of a number of different putative mammary oncogenes is postulated to account, in part, for this clonal selection.

5. These putative mammary oncogenes can be activated by insertion mutagenesis and chemically induced point mutations.

From the current database, we can postulate some of the general features of mouse mammary tumorigenesis. Any model must account for the multiplicity of putative mammary oncogenes and the observation that these genes are activated at variable frequencies in neoplasms from different strains. It is likely, although not yet proven, that several oncogenes may be expressed at the same time within cells of a single clonal lineage and that a given oncogene can be activated at different steps of neoplastic progression in different samples.

From both etiologic and morphogenic standpoints, it has been clear for some time that there are multiple "pathways" to mammary cancer (9,35). The recent molecular data confirm and extend these conclusions. Multiple molecular pathways exist, even for tumors with common etiologies and indistinguishable natural histories. The development of more probes for putative mammary oncogenes will likely reveal more intersample genetic variability, reflecting the complex mammary growth regulatory circuits and biochemical pathways that can be altered during tumorigenesis. From this vantage point, whether or not a given gene is active in the development of a given tumor is a function of the stochastic nature of the cell's evolution to autonomy. Mammary tumors share certain features, not because of a pro forma pathway to transformation, but because the mammary growth regulating mechanisms must serve as the substrates of activation, mutation, and selection in the cell progressing to malignancy.

ACKNOWLEDGMENTS

We thank our colleagues who critically read the manuscript for their many helpful comments. Work from our laboratory was supported by United States Public Health Services Grants 5 R01 CA 21454 and CA 30912 from the National Cancer Institute and by American Cancer Society Grant CD-235. In addition, we thank Debra Heras for preparing the manuscript.

REFERENCES

1. Aidells, B.D., and Daniel, C.W. (1974): Hormone-dependent mammary tumors in strain GR/A mice. I. Alternation between ductal and tumorous phases of growth during serial transplantation. *J. Natl. Cancer Inst.*, 52:1855–1863.
2. Ashley, R.L., Cardiff, R.D., and Fanning, T.G. (1980): Reevaluation of the effect of mouse mammary tumor virus infection on the BALB/c mouse hyperplastic outgrowth. *J. Natl. Cancer Inst.*, 65:977–986.
3. Ashley, R.H., Cardiff, R.D., Mitchell, D.J., Faulkin, L.J., and Lund, J.K. (1980): Development and characterization of mouse hyperplastic mammary outgrowth lines from BALB/cfC3H hyperplastic alveolar nodules. *Cancer Res.*, 40:4232–4242.
4. Bentvelzen, P. (1972): Hereditary infections with mammary tumor viruses in mice. In: *RNA Viruses and Host Genome in Oncogenesis,* edited by P. Emmelot, and P. Bentvelzen, pp. 309–337. North-Holland Publishing Company, Amsterdam.
5. Bentvelzen, P., and Hilgers, J. (1980): Murine mammary tumor virus. In: *Viral Oncology,* edited by G. Klein, pp. 331–355. Raven Press, New York.
6. Bern, H.A., and Nandi, S. (1961): Recent studies of the hormonal influence in mouse mammary tumorigenesis. In: *Progress in Experimental Tumor Research,* edited by F. Homburger, pp. 91–118. Hafner Publishing Company, New York.
7. Bishop, J.M. (1983): Cellular oncogenes and retroviruses. In: *Annual Review of Biochemistry,* Vol. 52, edited by E.E. Snell, P.D. Boyer, A. Meister, and C.C. Richardson, pp. 301–354. Annual Reviews, Palo Alto.
8. Bittner, J.J. (1936): Some possible effects of nursing on the mammary gland tumor incidence in mice. *Science,* 84:162.
9. Bittner, J.J. (1939): "Influences" of breast-cancer development in mice. *Public Health Rep.,* 54:1590–1597.
10. Blair, D.G., Cooper, C.S., Oskarsson, M.K., Eader, L.A., and Vande Woude, G.F. (1982): New method for detecting cellular transforming genes. *Science,* 218:1122–1125.
11. Cardiff, R.D. (1984): Protoneoplasia: The molecular biology of murine mammary hyperplasia. In: *Advances in Cancer Research,* Vol. 42, edited by G. Klein and S. Weinhouse, pp. 167–190. Academic Press, New York.
12. Cardiff, R.D., Fanning, T.G., Morris, D.W., Ashley, R.L., and Faulkin, L.J. (1981): Restric-

tion endonuclease studies of hyperplastic outgrowth lines from BALB/cfC3H mouse hyperplastic mammary nodules. *Cancer Res.,* 41:3024–3029.
13. Cardiff, R.D., Morris, D.W., and Young, L.J.T. (1983): Alterations of acquired mouse mammary tumor virus DNA during mammary tumorigenesis in BALB/cfC3H mice. *J. Natl. Cancer Inst.,* 71:1011–1019.
14. Cardiff, R.D., Wellings, S.R., and Faulkin, L.J. (1977): Biology of breast preneoplasia. *Cancer,* 39:2734–2746.
15. Cohen, J.C., Shank, P.R., Morris, V.L., Cardiff, R.D., and Varmus, H.E. (1979): Integration of the DNA of mouse mammary tumor virus in virus-infected normal and neoplastic tissue of the mouse. *Cell,* 16:333–345.
16. Cohen, J.C., and Varmus, H.E. (1980): Proviruses of mouse mammary tumor virus in normal and neoplastic tissues from GR and C3Hf mouse strains. *J. Virol.,* 35:298–305.
17. Cooper, G.M. (1982): Cellular transforming genes. *Science,* 218:801–806.
18. Dandekar S., Young, L.J.T., Cardiff, R.D., Zarbl, H., and Sukumar, S. (1986): Specific activation of the cellular Harvey-*ras* oncogene in dimethylbenzanthracene–induced mouse mammary tumors. *Mol. Cell Biol. (in press).*
19. Daniel, C.W., DeOme, K.B., Young, J.T., Blair, P.B., and Faulkin, L.J. (1968): The *in vivo* life span of normal and preneoplastic mouse mammary glands: A serial transplantation study. *Proc. Natl. Acad. Sci. U.S.A.,* 61:53–60.
20. DeOme, K.B., Faulkin, L.J., Jr., Bern, H.A., and Blair, P.B. (1959): Development of mammary tumors from hyperplastic alveolar nodules transplanted into gland-free mammary fat pads of female C3H mice. *Cancer Res.,* 19:515–526.
21. DeOme, K.B., and Medina, D. (1969): A new approach to mammary tumorigenesis in rodents. *Cancer,* 24:1255–1258.
22. DeOme, K.B. and Young, L.J.T. (1971): Hyperplastic lesions of the mouse and rat mammary glands. In: *Proceedings: Tenth International Cancer Research Congress,* Vol. 2, edited by R.E. Clair, R.W. Cumley, J.E. McCay and M.M. Copeland, pp. 474–483. Yearbook Medical Publishers, Chicago.
23. Dickson, C., Smith, R., Brookes, S., and Peters, G. (1984): Tumorigenesis by mouse mammary tumor virus: Proviral activation of a cellular gene in the common integration region *int-2*. *Cell,* 37:529–536.
24. Ehmann, U.K., Peterson, Jr., W.D., and Misfeldt, D.S. (1984): To grow mouse mammary epithelial cells in culture. *J. Cell Biol.,* 98:1026–1032.
25. Etkind, P.R., and Sarkar, N.H. (1983): Integration of new endogenous mouse mammary tumor virus proviral DNA at common sites in the DNA of mammary tumors of C3Hf mice and hypomethylation of the endogenous mouse mammary tumor virus proviral DNA in C3Hf mammary tumors and spleens. *J. Virol.,* 45:114–123.
26. Fanning, T.G., and Cardiff, R.D. (1984): Alterations of mouse mammary tumor virus DNA during mammary tumorigenesis. In: *Advances in Viral Oncology,* Vol. 4., edited by G. Klein, pp. 71–94. Raven Press, New York.
27. Fanning, T.G., Puma, J.P., and Cardiff, R.D. (1980): Selective amplification of mouse mammary tumor virus in mammary tumors of GR mice. *J. Virol.,* 36:109–114.
28. Fanning, T.G., Vassos, A.B., and Cardiff, R.D. (1982): Methylation and amplification of mouse mammary tumor virus DNA in normal, premalignant, and malignant cells of GR/A mice. *J. Virol.,* 41:1007–1013.
29. Fasano, O., Birnbaum, D., Edlund, L., Fogh, J., and Wigler, M. (1984): New human transforming genes detected by a tumorigenicity assay. *Mol. Cell. Biol.,* 9:1695–1705.
30. Faulkin, L.J., Jr., and DeOme, K.B. (1960): Regulation of growth and spacing of gland elements in the mammary fat pad of the C3H mouse. *J. Natl. Cancer Inst.,* 24:953–969.
31. Foulds, L. (1956): The histologic analysis of mammary tumors of mice. II. The histology of responsiveness and progression. The origins of tumors. *J. Natl. Cancer Inst.,* 17:713–753.
32. Foulds, L. (1969): *Neoplastic Development.* Academic Press, New York.
33. Graham, K.A., Richardson, C.L., Minden, M.D., Trent, J.M., and Buick, R.N. (1985): Varying degrees of amplification of the N-*ras* oncogene in the human breast cancer cell line MCF-7. *Cancer Res.,* 45:2201–2205.
34. Groner, B., and Hynes, N.E. (1980): Number and location of mouse mammary tumor virus proviral DNA in mouse DNA of normal tissue and of mammary tumors. *J. Virol.,* 33:1013–1025.
35. Haaland, M. (1911): Spontaneous tumours in mice. *Sci. Rep. Imperial Cancer Res. Fund,* 4:1–113.

36. Hayward, W.S., Neel, B.G., and Astrin, S.M. (1981): Activation of a cellular *onc* gene by promoter insertion in ALV-induced lymphoid leukosis. *Nature*, 290:475–480.
37. Hayward, W.S., Neel, B.G., and Astrin, S.M. (1982): Avian leukosis viruses: Activation of cellular "oncogenes." In: *Advances in Viral Oncology*, Vol. 1, edited by G. Klein, pp. 207–231. Raven Press, New York.
38. Heppner, G.H. (1984): Tumor heterogeneity. *Cancer Res.*, 44:2259–2265.
39. Heston, W.E., and Vlahakis, G. (1971): Mammary tumors, plaques, and hyperplastic alveolar nodules in various combinations of mouse inbred strains and the different lines of the mammary tumor virus. *Int. J. Cancer*, 7:141–148.
40. Hynes, N.E., Jaggi, R., Kozma, S.C., Ball, R., Muellener, D., Wetherall, N.T., Davis, B.W., and Groner, B. (1985): New acceptor cell for transfected genomic DNA: Oncogene transfer into a mouse mammary epithelial cell line. *Mol. Cell. Biol.*, 5:268–272.
41. Imai, S., Tsubura, Y., and Hilgers, J. (1984): Urethan-induced mammary tumorigenesis in a murine mammary tumor virus (MuMTV)-positive mouse strain: Evidence for a keratinized nodule as an MuMTV-negative precursor lesion for squamous cell tumors. *J. Natl. Cancer Inst.*, 73:935–941.
42. King, C.R., Kraus, M.H., and Aaronson, S.A. (1985): Amplification of a novel v-*erb*B-related gene in a human mammary carcinoma. *Science*, 229:974–976.
43. Klein, G., and Klein, E. (1985): Evolution of tumours and the impact of molecular oncology. *Nature*, 315:190–195.
44. Kozbor, D., and Croce, C.M. (1984): Amplification of the c-*myc* oncogene in one of five human breast carcinoma cell lines. *Cancer Res.*, 44:438–441.
45. Kraus, M.H., Yuasa, Y., and Aaronson, S.A. (1984): A position 12-activated H-*ras* oncogene in all HS578T mammary carcinosarcoma cells but not normal mammary cells of the same patient. *Proc. Natl. Acad. Sci. U.S.A.*, 81:5384–5388.
46. Lane, M.-A., Sainten, A., and Cooper, G.M. (1981): Activation of related transforming genes in mouse and human mammary carcinomas. *Proc. Natl. Acad. Sci. U.S.A.*, 78:5185–5189.
47. MacInnes, J.I., Chan, E.C.M.L., Percy, D.H., and Morris, V.L. (1981): Mammary tumors from GR mice contain more than one population of mouse mammary tumor virus-infected cells. *Virology*, 113:119–129.
48. Medina, D. (1973): Preneoplastic lesions in mouse mammary tumorigenesis. In: *Methods in Cancer Research*, Vol. 7, edited by H. Busch, pp. 3–53. Academic Press, New York.
49. Medina, D. (1976): Mammary tumorigenesis in chemical carcinogen-treated mice. VI. Tumor-producing capabilities of mammary dysplasias in BALB/cCrgl mice. *J. Natl. Cancer Inst.*, 57:1185–1189.
50. Medina, D. (1976): Preneoplastic lesions in murine mammary cancer. *Cancer Res.*, 36:2589–2595.
51. Medina, D. (1978): Preneoplasia in breast cancer. In: *Breast Cancer*, Vol. 2, edited by W.L. McGuire, pp. 47–102. Plenum Medical Book Company, New York.
52. Medina, D., and DeOme, K.B. (1968): Influence of mammary tumor virus on the tumor-producing capabilities of nodule outgrowth free of mammary tumor virus. *J. Natl. Cancer Inst.*, 40:1303–1308.
53. Michalides, R., Wagenaar, E., and Sluyser, M. (1982): Mammary tumor virus DNA as a marker for genotypic variance within hormone-responsive GR mouse mammary tumors. *Cancer Res.*, 42:1154–1158.
54. Modjtahedi, N., Lavialle, C., Poupon, M.-F., Landin, R.-M., Cassingena, R., Monier, R., and Brison, O. (1985): Increased level of amplification of the c-*myc* oncogene in tumors induced in nude mice by a human breast carcinoma cell line. *Cancer Res.*, 45:4372–4379.
55. Nandi, S., and McGrath, C.M. (1973): Mammary neoplasia in mice. In: *Advances in Cancer Research*, Vol. 17, edited by G. Klein, S. Weinshouse, and A. Haddow, pp. 353–414. Academic Press, New York.
56. Neel, B.G., Hayward, W.S., Robinson, H.L., Fang, J., and Astrin, S.M. (1981): Avian leukosis virus-induced tumors have common proviral integration sites and synthesize discrete new RNAs: Oncogenesis by promoter insertion. *Cell*, 23:323–334.
57. Nusse, R., van Ooyen, A., Cox, D., Fung, Y.K.T., and Varmus, H. (1984): Mode of proviral activation of a putative mammary oncogene *int*-1 on mouse chromosome 15. *Nature*, 307:131–136.
58. Nusse, R., and Varmus, H.E., (1982): Many tumors induced by the mouse mammary tumor virus contain a provirus integrated in the same region of the host genome. *Cell*, 31:99–109.

59. Payne, G.S., Bishop, J.M., and Varmus, H.E. (1982): Multiple arrangements of viral DNA and an activated host oncogene in bursal lymphomas. *Nature,* 295:209–214.
60. Payne, G.S., Courtneidge, S.A., Crittenden, L.B., Fadly, A.M., Bishop, J.M., and Varmus, H.E. (1981): Analysis of avian leukosis virus DNA and RNA in bursal tumors: Viral gene expression is not required for maintenance of the tumor state. *Cell,* 23:311–322.
61. Peters, G., Brookes, S., Smith, R., and Dickson, C. (1983): Tumorigenesis by mouse mammary tumor virus: Evidence for a common region for provirus integration in mammary tumors. *Cell,* 33:369–377.
62. Peters, G., Kozak, C., and Dickson, C. (1984): Mouse mammary tumor virus integration regions *int*-1 and *int*-2 map on different mouse chromosomes. *Mol. Cell. Biol.,* 4:375–378.
63. Peters, G., Lee, A.E., and Dickson, C. (1984): Activation of cellular gene by mouse mammary tumour virus may occur early in mammary tumour development. *Nature,* 309:273–275.
64. Ponta, H., Kennedy, N., Skroch, P., Hynes, N.E., and Groner, B. (1985): Hormonal response region in the mouse mammary tumor virus long terminal repeat can be dissociated from the proviral promoter and has enhancer properties. *Proc. Natl. Acad. Sci U.S.A.* 82:1020–1024.
65. Ringold, G.M. (1983): Regulation of mouse mammary tumor virus gene expression by glucocorticoid hormones. In: *Current Topics in Microbiology and Immunology,* Vol. 106, edited by P.K. Vogt and H. Koprowski, pp. 79–103. Springer-Verlag, New York.
66. Rubin, H. (1985): Cancer as a dynamic developmental disorder. *Cancer Res.,* 45:2935–2942.
67. Shih, C., Shilo, B.-Z., Goldfarb, M.P., Dannenberg, A., and Weinberg, R.A. (1979): Passage of phenotypes of chemically transformed cells via transfection of DNA and chromatin. *Proc. Natl. Acad. Sci. U.S.A.,* 76:5714–5718.
68. Slagle, B.L., Wheeler, D.A., Hager, G.L., Medina, D., and Butel, J.S. (1985): Molecular basis of altered mouse mammary tumor virus expression in the D-2 hyperplastic alveolar nodule line of BALB/c mice. *Virology,* 143:1–15.
69. Sluyser, M., and van Nie, R. (1974): Estrogen receptor content and hormone-responsive growth of mouse mammary tumors. *Cancer Res.,* 34:3253–3257.
70. Staff, Jackson Memorial Laboratory (1933): The existence of nonchromosomal influence in the incidence of mammary tumors in mice. *Science,* 78:465–466.
71. Stewart, T.A., Pattengale, P.K., and Leder, P. (1984): Spontaneous mammary adenocarcinomas in transgenic mice that carry and express MTV/*myc* fusion genes. *Cell,* 38:627–637.
72. Sukumar, S., Notario, V., Martin-Zanca, D., and Barbacid M. (1983): Induction of mammary carcinomas in rats by nitroso-methylurea involves malignant activation of H-*ras*-1 locus by single point mutations. *Nature,* 306:658–661.
73. van Nie, R., and Dux, A. (1971): Biological and morphological characteristics of mammary tumors in GR mice. *J. Natl. Cancer Inst.,* 46:885–897.
74. van Ooyen, A., and Nusse, R. (1984): Structure and nucleotide sequence of the putative mammary oncogene *int-1:* Proviral insertions leave the protein-encoding domain intact. *Cell,* 39:233–240.
75. Varmus, H.E. (1982): Recent evidence for oncogenesis by insertion mutagenesis and gene activation. *Cancer Surveys,* 1:309–319.
76. Varmus, H.E. (1984): The molecular genetics of cellular oncogenes. In: *Annual Review of Genetics,* Vol. 18, edited by H.L. Roman, A. Campbell, and L.M. Sandler, pp. 553–612. Annual Reviews. Palo Alto.
77. Visscher, M.B., Green, R.G., and Bittner, J.J. (1942): Characterization of milk influence in spontaneous mammary carcinoma. *Proc. Soc. Exp. Biol. Med.,* 49:94–96.
78. Voyles, B.A., and McGrath, C.M. (1976): Markers to distinguish normal and neoplastic mammary epithelial cells *in vitro:* Comparison of saturation density, morphology, and concanavalin A reactivity. *Int. J. Cancer,* 18:498–509.
79. Weinberg, R.A. (1982): Oncogenes of spontaneous and chemically induced tumors. In: *Advances in Cancer Research,* Vol. 36 edited by G. Klein, and S. Weinhouse, pp. 149–163. Academic Press, New York.
80. Zarbl, H., Sukumar, S., Arthur, A.V., Martin-Zanca, D., and Barbacid, M. (1985): Direct mutagenesis of Ha-*ras*-1 oncogenes by *N*-nitroso-*N*-methylurea during initiation of mammary carcinogenesis in rats. *Nature,* 315:382–385.

Role of Oncogene Amplification in Tumor Progression

Yoichi Taya, Masaaki Terada, and Takashi Sugimura

National Cancer Center Research Institute, Tsukiji, Chuo-ku, Tokyo 104, Japan

It is now established that viral oncogenes in acute transforming retroviruses are derived from proto-oncogenes by transduction and are responsible for the transforming activity of the viruses. More than 20 proto-oncogenes have been identified, and most of these have homologous sequences with those in viral oncogenes. In addition, some cellular oncogenes that do not have viral oncogene counterparts have been isolated from human and animal cancers by transfection assay (5,14, 27,30–32).

Activation of proto-oncogenes occurs by at least the following types of mechanisms: insertion of a promoter or enhancer element, translocation, point mutation of a structural gene, and amplification. Amplification of proto-oncogenes is one of the mechanisms by which proto-oncogene products are increased. The associations of amplification of different proto-oncogenes with different types of malignancies in human and animals have been reported (20,67,80). In this chapter, studies on amplification of proto-oncogenes are reviewed with emphasis on the role of this amplification in the progression of cancer.

One characteristic of malignant cells ia aneuploidy. Double-minute chromosomes (DM) and homogeneously staining regions (HSR) in chromosomes are often found in cancers (4). Furthermore, an increased DNA content per cell is frequently observed in cancer cells when the DNA content is measured by cytochemical staining (3). Accordingly, studies on amplifications of cellular genes other than proto-oncogenes in cancers are also reviewed.

AMPLIFICATION OF PROTO-ONCOGENES

The oncogene amplifications that have been found in human and animal tumor cells are listed in Table 1. Specific properties of each of these amplified proto-oncogenes are described in the following section.

TABLE 1. *List of amplified proto-oncogenes*

Cell line or tumor (human origin)	Proto-oncogene	Degree of amplification (fold)	Ref.
HL-60 (acute promyelocytic leukemia)	myc	20	9, 13, 55
Stomach cancers	myc	8–30	33, 50, 77
SKBR-3 (breast carcinoma)	myc	10	37
COLO320 (colon carcinoma)	myc	30	1
SEWA (osteosarcoma)[a]	myc	15–30	71
Morris hepatoma 7794A[b]	myc	5–10	25
SW613-S (breast carcinoma)	myc	5–90	46
Colon tumor 36[a]	myc	30–40	88
Small cell lung cancer	myc N-myc L-myc	20–76	42, 52, 66
Neuroblastomas	N-myc	25–700	35, 48, 68
Retinoblastomas	N-myc	10–200	39
ML-1, 2, 3 (acute myelogenous leukemia)	myb	5–10	58
COLO201, 205 (colon carcinoma)	myb	10	2
KT883 (epidermoid lung carcinoma)	Ki-ras	10–20	45
MCF-7 (breast carcinoma)	N-ras	20	16, 22
Y1 (adrenocortical tumor)[a]	Ki-ras	50	21, 69
Ya (bladder carcinoma)	Ha-ras	5–10	24
ABT165 (bladder carcinoma)	Ki-ras	40	19
SK-2 (melanoma)	Ha-ras	10–20	75
Adenocarcinoma of the ovary	Ki-ras	10–20	18
Glioblastomas	erbB-1	6–60	40
A431 (epidermoid carcinoma)	erbB-1	15–20	41, 44, 83
Squamous carcinomas	erbB-1	2–50	86
K562 (chronic myelogenous leukemia)	abl	10	10, 76
Stomach cancer	yes	4–6	74
Lu-65 (giant-cell lung carcinoma)	myc Ki-ras	8 10	81
Pancreatic cancer	myc Ki-ras	50 6	

[a]Mouse
[b]Rat

c-*myc*

The c-*myc* gene is the cellular homologue of the transforming gene carried by the myelocytomatosis virus. Expression of c-*myc* is shown to be under growth signal control in lymphocytes, fibroblasts (29), and regenerating rat liver (43). The c-*myc* gene apparently plays an important role in transition of the cells from G_0 to G_1 of the cell cycle.

Amplification of c-*myc* has been observed in several cancer cell lines and transplantable tumors derived from cancer tissues. These include HL-60, an acute promyelocytic leukemia cell line (9,13), a colon carcinoma cell line (COLO320) (1), small-cell lung cancer cell lines (42,52), a xenograft of a breast carcinoma (SKBR-3) (37), a xenograft of a giant-cell carcinoma of the lung (Lu-65) (81), and transplantable stomach cancers (50,77). A mouse osteosarcoma cell line (SEWA) (71), Morris hepatoma 7794A (25), and a chemically induced mouse colon tumor (88 were also reported to have amplified c-*myc*. COLO320 cell lines, derived from a colon carcinoma of endocrine origin were found to have DM or HSR containing amplified c-*myc* sequences in this X chromosome. Since the chromosomal localization of c-*myc* is $8q24$ in normal human cells, the c-*myc* sequences were moved to new chromosomal sites (1).

Although amplification of c-*myc* has mainly been observed in cell lines or tumors transplanted into nude mice, it does occur *in vivo;* c-*myc* amplification was observed in uncultured promyelocytic leukemic cells before establishment of HL60 cells (13) and in stomach cancer (33) and pancreatic adenocarcinoma (85a). Only a few solid tumors show c-*myc* amplification *in vivo*. No amplification of c-*myc* was observed in 31 samples of DNAs from stomach cancers *in vivo*, whereas two of 14 samples of DNAs from stomach cancers xenografted into nude mice contained amplified c-*myc* sequences (50). It is possible that cancer cells with c-*myc* amplification *in vivo* cannot be detected because of the heterogeneity of the cancer cells *in vivo* and that these cells with c-*myc* amplification have a selective growth advantage in culture or in nude mice. It is also possible that amplification of c-*myc* might have occurred during passages in culture *in vitro* or in nude mice.

Amplification of c-*myc* does not necessarily lead to increased c-*myc* mRNA. In one stomach cancer, no increased c-*myc* mRNA was observed despite the presence of amplified c-*myc* (77). In HL60 cells, c-*myc* was amplified about 30-fold with corresponding increase in c-*myc* mRNA. On induction of differentiation *in vitro*, c-*myc* mRNA was decreased markedly, whereas the copy number of c-*myc* remained unchanged (13).

HL-60 cells contain an activated N-*ras* gene in addition to amplified c-*myc* (49). The Lu-65 tumor (81) and a pancreatic adenocarcinoma with amplified c-*myc* contained activated K-*ras* with a point mutation (85a). These results support the idea that carcinogenesis is a multistep process and further strengthen the concept that at least two cooperating genes are required for transformation of normal cells.

N-*myc*

Amplification of the N-*myc* gene is frequently observed in human neuroblastoma cell lines and neuroblastoma tumors (35,48,68). It has also been found in some human retinoblastomas (39) and in small-cell lung cancers (52). Since small-cell lung cancers have many properties in common to neuroendocrine tumors, N-*myc* amplification may be specifically related to neuroendocrine tumors.

High levels of N-*myc* mRNA are present in all cancers that have an amplified N-*myc* gene (34,39,70). Appreciable amounts of this RNA are present in a number of cell lines derived from other human tumors or in normal cells. In contrast, readily detectable levels of N-*myc* mRNA are found in human and murine neuroblastoma cell lines and in retinoblastomas in which this gene is not amplified (34,39,70).

The N-*myc* gene is normally localized on human chromosome 2*p*23-24 (73). However, amplified N-*myc* genes are localized in HSR on different chromosomes in many human neuroblastomas (15,73), suggesting the absence of a preferred chromosomal site for N-*myc* integration and amplification.

c-*myb*

Amplification of the c-*myb* gene has been found in cell lines ML-1, ML-2, and ML-3 (58), which were separately cultured from cells of a patient with acute myelogenous leukemia, and in two cell lines (COLO201 and COLO205) derived independently from a single adenocarcinoma of human colon (2). These cells contained high levels of *myb* mRNA. The *myb* gene has structural and possibly functional similarities to the *myc* gene (61), is specifically expressed in hematopoietic cells (85), and appears to be correlated with the stage of differentiation of these cells (85).

ras Gene Family

Amplification of *ras* gene family has been found in a variety of cancers; however, the c-Ki-*ras* gene is most frequently amplified of these three *ras* genes. In a giant-cell carcinoma of the lung, Lu-65 (54a,81), a melanoma SK-2 (75), and a pancreatic cancer (85a), amplification of the *ras* genes was found to be associated with point mutational activation; in all three cases, it was the point-mutated allele that was amplified. Conversely, normal N-*ras* gene and c-Ki-*ras* were amplified in a breast carcinoma MCF-7 (16) and a bladder cancer ABT165 (19), respectively.

Ligation of a transcriptional enhancer to a proto-oncogene is expected to have a similar effect to proto-oncogene amplification because high levels of proto-oncogene proteins are produced in both cases. In this regard, it is noteworthy that Chang *et al.* (8) made the interesting observation that a normal c-Ha-*ras* gene linked to a long terminal repeat (LTR) induces tumorigenic transformation of NIH3T3 cells. Fasano *et al.* (16) also observed the tumorigenic activity of an amplified normal N-*ras* gene using a sensitive bioassay for transforming genes based on the tumorigenicity of cotransfected NIH3T3 cells in nude mice, although no transforming activity was detected by ordinary NIH3T3 cells focus assay. The finding of Spandidos and Wilkie (79) is very interesting; they showed that when linked to transcriptional enhancers, the point-mutated Ha-*ras* gene induced complete malignant transformation of early passage cells, whereas the normal Ha-*ras* gene induced only immortalization.

c-*erb*B

The proto-oncogene c-*erb*B-1, which is considered to be indentical to the gene for the epidermal growth factor (EGF) receptor, is amplified in several squamous carcinomas (41,44,83,86) and glioblastomas (40). Overexpression of EGF receptors on such cells is also reported. Abnormal *erb*B transcripts that encoded truncated external domains were detected in addition to normal *erb*B mRNA in an epidermoid carcinoma cell line A431 and in a glioblastoma, suggesting rearrangement among the amplified *erb*B genes. These truncations are different from that of v-*erb*B in which the EGF binding site is deleted. Further analysis is needed to determine the interrelations of gene structure, aberrant mRNA expression, and malignant transformation.

c-*abl*

Chronic myelogenous leukemia (CML) is a pluripotent stem cell disease characterized in more than 90% of cases by the presence of the Philadelphia chromosome. This chromosome is generated by a reciprocal exchange between chromosome 9, which bears the c-*abl* gene, and chromosome 22. Amplification of the *abl* gene was found in one CML line, K562 (10,76). It was also observed that the λ light-chain constant region immunoglobulin genes, which were normally located on chromosome 22, were amplified in K562 (11). A novel *abl* mRNA of larger size is produced in CML cells, including K562. This abnormal mRNA has been shown to be formed as a result of fusion of a *bcr* gene on chromosome 22 to *abl* gene on chromosome 9, producing an *abl* polypeptide with an altered amino-terminal segment (26,78). This altered c-*abl* protein has tyrosine kinase activity like the v-*abl* protein, whereas the normal c-*abl* protein does not have this activity (36). Thus, the abnormal c-*abl* protein with tyrosine kinase activity is overproduced in K562 cells. The *bcr* gene is localized on the same band on chromosome 22 as the λ light-chain constant region genes, but it nevertheless seems to be a different gene.

Two Proto-Oncogene Amplification

The amplifications of two oncogenes, c-*myc* and Ki-*ras,* were found in a human lung giant-cell carcinoma, Lu-65 (54a,81) and in primary and metastatic pancreatic cancers in a patient (85a). In both cases, point mutations were found in position 12 of the c-Ki-*ras* gene. Therefore, point mutational activation is accompanied by amplifications of c-*myc* and the c-Ki-*ras* genes. Further analysis showed that the mutated allele of the Ki-*ras* gene was amplified in both cancers. The same complicated activations were found in the primary pancreatic tumor and its metastatic tumor. By contrast, no activation was found in normal cells of the same patient. Accordingly, at least in these pancreatic cancers, these multiple activations are not artifacts that occur during transplantation into nude mice or during culture. Coexistence of the amplified c-*myc* gene with an N-*ras* gene activated by a point mutation is also reported in the HL-60 human promyelocytic leukemia cell line (49). These results are consistent with the findings that at least two transforming genes are required for malignant transformation of primary fibroblasts (38,53,65) and with the multistep process of carcinogenesis.

DOUBLE-MINUTE BODIES, HOMOGENEOUSLY STAINING REGIONS, AND AMPLIFIED GENES OTHER THAN PROTO-ONCOGENES IN CANCERS

Double-minute bodies (DM) are small extra-chromosomal nuclear entities of 0.3 to 0.5 μm diameter with no centromeres that segregate unpredictably at mitosis. Homogenously staining regions (HSR) of centromeric chromosomes are chromosomal regions that do not show characteristic differential staining, but stain uniformly (14,67,80). DM and HSR are frequently seen in cancer cells. Much information on the molecular nature and origin of DM and HSR has been obtained from studies on cultured cells selected for drug resistance. For example, Chinese hamster cell lines obtained by stepwise selection for methotrexate resistance had increased dihydrofolate reductase, which is the target enzyme for methotrexate. These cells contained an amplified dihydrofolate reductase gene, which was localized in DM or HSR. DM are apparently unstable forms of amplified DNA sequences and could possibly be the precursors of HSR, which are stable forms of amplified DNA sequences. For more detailed information on the nature and mechanisms involved in gene amplification in drug resistance, the reader is referred to reviews on the subject (56,67,80).

In several studies in which amplified proto-oncogenes were found, amplified proto-oncogene sequences were shown to be localized within DM or HSR. The neuroblastoma cell line IMR-32 contains two larger chromosome 1 with an HSR on the short arm (1*p*3) (35,73). Using fluorescence-activated flow sorting of metaphase chromosomes, HSR-chromosomes were enriched: a DNA library was then constructed from them, and sequences amplified 50-fold in these cells were cloned. These amplified sequences of 3,000 kilo-base-pairs (kbp) were found to be localized in the HSR of chromosome 1 and short arm of chromosome 2 in IMR-32, whereas they were localized in the short arm of chromosome 2 in normal cells. Amplification of these sequences in IMR-32 cells was suggested to be due to transposition from chromosome 2 to chromosome 1. Results showed that, at least in IMR-32 cells, amplification involved not only relocation of DNA from specific genomic domains but also the formation of novel units by splicing together of very distant fragments. Furthermore, one of the amplified sequences was shown to contain the proto-oncogene N-*myc*. It should be noted that not all the sequences amplified in IMR-32 are amplified in the other neuroblastomas. Amplification of N-*myc* was established to be a common event in untreated human neuroblastomas and to be closely correlated with advanced stages of diseases, as described in a separate section of this chapter. N-*myc* amplification was not restricted to neuroblastomas; it was also observed in retinoblastomas and small-cell lung carcinomas.

A mouse tumor cell line, SEWA, derived from an osteosarcoma induced by infection with polyoma virus, was shown to contain multiple DM when the cells were grown in a mouse, but the number of DM decreased to an undetectable level when the cells were grown in culture. The DM of SEWA cells were thought to contain genes involved in malignancy. Recently, the cells were found to contain amplified c-*myc* localized in the DM. It is possible that enhancement of expression of c-*myc* after amplification confers on the host cells a selective advantage for tumor formation in the natural host (71). Other examples of amplified genes associated with DM or HSR are c-*myc* amplification in COLO 320 human carcinoma cells (1) and c-Ki-*ras* amplification in Y1 cells (69).

In solid tumors, analysis of cytogenetic abnormalities has been hampered by the technical difficulty of obtaining enough metaphase chromosomes. However, by recently developed methods for analysis of the DNA content per cell by cytofluorometry, flow cytometry, and cytophotometry, abnormalities in DNA contents, including hyperploidy, have been shown to be frequent in malignant tumors, especially undifferentiated types (3). For determination of the DNA sequences amplified, the DNA renaturation method in gel has been used to analyze DNAs from a neuroblastoma, a retinoblastoma, and a small-cell lung carcinoma (51). This method was first developed for analyzing amplified sequences in cells that were resistant to anticancer drugs (63,64). In this method, DNA was digested completely by a restriction enzyme, a portion of which was labeled with α-^{32}P-dCTP by T_4-DNA polymerase. The labeled DNA fragments were mixed with unlabeled DNA digested with the same restriction enzyme and subjected to electrophoresis in agarose. DNA denaturation and renaturation were performed in the gel, followed by S_1 nuclease digestion of unannealed DNA fragments. Results by this method showed that a neuroblastoma cell line, a retinoblastoma cell line, and a small-cell lung carcinoma cell line all contained more than 200-fold amplified sequences other than N-*myc*. The sizes of amplified sequences were estimated to be 1,000 to 3,000 kbp. Since the amplified sequences can be cloned directly from the gel, this simple method appears to have potential value for analysis of amplified sequences in other cancers. It is assumed, and in some cases proved, that DNA sequences other than cellular proto-oncogenes are amplified, but it is still unknown whether these amplified sequences play roles in malignant phenotypes. Studies on the sequences amplified in cancer cells will answer these questions.

PROTO-ONCOGENE AMPLIFICATION AND TUMOR PROGRESSION

Evidence for a correlation between oncogene amplification and tumor progression has been obtained in the cases of the N-*myc* and *myc* genes. Amplification of N-*myc* in neuroblastoma was found to be closely correlated with advanced stages of the disease and with the ability of cells to grow *in vitro* as an established cell line. Amplification of N-*myc* was not found in 15 patients in stage 1 or 2, but was found in 24 of 48 patients in stage 3 or 4 (6). In human small-cell lung cancers, amplification of the c-*myc*, N-*myc*, or L-*myc* gene was frequently observed in cells of variant classes, which are considered to have more malignant phenotype than classic small-cell lung cancers (52). For more detailed information on the relationship between malignant phenotypes of small-cell lung cancers and amplifications of *myc* genes, the reader should refer to the chapter by B. J. Brooks, et al., *this volume*.

The proteins of these *myc* family proto-oncogenes, as well as the *myc* gene, are presumably localized in the nucleus, although no N-*myc* or L-*myc* protein has yet been detected. It is noteworthy that of 20 proto-oncogenes, these proto-oncogenes are the most frequently amplified. Therefore, it is important to elucidate the physiological functions of these proto-oncogene proteins, which at present are uncertain. The proteins of c-*myc* and c-*myb* seem to be DNA-binding proteins (47,59). The N-*myc* and L-*myc* proteins should have similar properties. It has been suggested that the c-*myc* protein acts as a competence factor in the cell cycle to promote the progression of cells to the S phase, because when injected into the nuclei of quiescent cells, it stimulated DNA synthesis (28). The involvement of *myc* gene in the cell cycle was first suggested by Kelly *et al.* (29), showing that *myc* gene was induced in quiescent cells by platelet-derived growth factor or mitogen. The findings that the N-*myc* and c-*myb* genes, like the c-*myc* gene, can cooperate with the activated *ras* gene to transform culture cells at early passages signify that these genes can replace the *myc*-specific function (57,72,87).

The expressions of these genes seem to depend on the type of cells and their stage of differentiation or development, at least in several types of cells and tissues. In this respect, a variety of lines of evidence have been accumulated on the *myc* gene. First, Croce and collaborators (12,17,54) reported that in somatic cell hybrids between Burkitt lymphoma cells and other cells, expression of the c-*myc* gene depends on the stage of B-cell differentiation. Second, a cell-type-specific pattern of c-*myc* gene expression was shown in developing human embryos using *in situ* hybridization (60). Third, marked decrease in expression of the c-*myc* gene is observed on induction of differentiation of HL-60 or teratocarcinoma cells (7,62,85). A similar decrease in expression of N-*myc* was observed during retinoic acid-induced differentiation of a neuroblastoma (82).

In cancer cells that have an amplified *myc* family or c-*myb* gene, the amplified proto-oncogene is presumably expressed effectively by the presence of factors needed for its expression. Expression of such a gene in cancer cells will decrease after induction of differentiation if the factor dissappears, even though the proto-oncogene is still amplified. It is therefore assumed that amplified proto-oncogene should exert its biological effects at a specific stage of differentiation or development of the cells.

Another proto-oncogene, c-Ki-*ras*, was found to have a point mutation at the twelfth codon in a pancreatic cancer, and this mutated c-Ki-*ras* was subsequently amplified, indicating that amplification occurred during progression of the cancer (85a).

Interestingly, DNA damaging agents, including ultraviolet rays and N-acetoxy-N-acetylaminofluorene enhanced amplification of the dihydrofolate reductase gene (67). Not only carcinogenic agents, but a tumor promoter, 12-O-tetradecanoyl-phorbol-13-acetate (TPA), induced amplification of the folic acid reductase gene (84), presumably by increasing misfiring of replication. It was also shown that potent tumor promoters of different classes, TPA, and dihydroteleocidin B induced amplification of methallothionein 1 gene (23). It is possible that similar types of mechanisms might be involved in amplification of proto-oncogenes during the tumor-promoting stage, resulting in growth advantage of the cells with amplified proto-oncogenes. No experimental system is available for induction of amplification of a specific proto-oncogene, and the mechanisms involved in amplification of proto-oncogenes are still unknown.

CONCLUDING REMARKS

Amplification of DNA sequences in malignant cells involves not only increase in DNA sequences but also translocation and integration of sequences. Studies on amplified sequences including proto-oncogene might lead to a better understanding of the molecular basis for chromosomal instability or gene rearrangement in malignant cells and the molecular mechanisms that keep normal cells in the diploid state.

Amplification of proto-oncogenes in many instances resulted in overproduction of the products. Elucidation of the functional roles of these proto-oncogene products in normal cells and malignant cells will certainly lead to a better understanding of the significance of the amplification of proto-oncogenes during the carcinogenic process, including tumor progression, and of molecular mechanisms involved in gene amplification. Studies on proto-oncogene amplification are important not only from the molecular biological point of view, but also from the clinical point, as exemplified in the case of N-*myc* amplification in neuroblstoma. It is possible that advanced clinical stages of other types of cancers might be associated with amplification of specific proto-oncogenes.

REFERENCES

1. Alitalo, K., Schwab, M., Lin, C.C., Varmus, H.E., and Bishop, J.M. (1983): Homogeneously staining chromosomal regions contain amplified copies of an abundantly expressed cellular oncogene (c-*myc*) in malignant neuroendocrine cells from a human colon carcinoma. *Proc. Natl. Acad. Sci. U.S.A.*, 80:1707–1711.
2. Alitalo, K., Winqvist, R., Lin, C.C., de la Chapelle, A., Schwab, M., and Bishop, J.M. (1984): Aberrant expression of an amplified c-*myb* oncogene in two cell lines from a colon carcinoma. *Proc. Natl. Acad. Sci. U.S.A.*, 81:4534:4538.
3. Barlogie, B., Brewinko, B., Schumann, J., Göhde, W., Dosik, G., Latreille, J., Johnston, D.A., and Freireich, E.J. (1980): Cellular DNA content as a marker of neoplasia in man. *Am. J. Med.*, 69:195–203.

4. Biedler, J.L., Melera, P.W., ans Spengler, B.A. (1983): Chromosome abnormalities and gene amplification: Comparison of antifolate-resistant and human neuroblastoma cell systems. In: *Chromosomes and Cancer,* edited by J.D. Rowley and J.E. Ultmann, pp. 117–138. Academic Press, Inc., New York.
5. Bishop, J.M. (1983): Cellular oncogenes and retroviruses. *Ann. Rev. Biochem.,* 52:301–354.
6. Brodeur, G.M., Seeger, R.C., Schwab, M., Varmus, H.E., and Bishop, J.M. (1984): Amplification of N-*myc* in untreated human neuroblastomas correlates with advanced disease stage. *Science,* 224:1121–1124.
7. Campisi, J., Gray, H.E., Pardee, A.B., Dean M., and Sonenshein, G.E. (1984): Cell-cycle control of c-*myc* but not c-*ras* expression is lost following chemical transformation. *Cell,* 36:241–247.
8. Chang, E.H., Furth, M.E., Scolnick, E.M., and Lowy, D.R. (1982): Tumorigenic transformation of mammalian cells induced by a normal human gene homologous to the oncogene of Harvey murine sarcoma virus. *Nature,* 297:479–483.
9. Collins, S., and Groudine, M. (1982): Amplification of endogenous *myc*-related DNA sequences in a human myeloid leukemia cell line, *Nature,* 298:679–681.
10. Collins, S.J., and Groudine, M.T. (1983): Rearrangement and amplification of c-*abl* sequences in the human chronic myelogenous leukemia cell line K-562. *Proc. Natl. Acad. Sci. U.S.A.,* 80:4813–4817.
11. Collins, S.J., Kubonishi, I., Miyoshi, I., and Groudine, M.T. (1984): Altered transcription of the c-*abl* oncogene in K-562 and other chronic myelogenous leukemia cells. *Science,* 225:72–74.
12. Croce, C.M., Erikson, J., ar-Rushdi, A., Aden, D., and Nishikura, K. (1984): Translocated c-*myc* oncogene of Burkitt lymphoma is transcribed in plasma cells and repressed in lymphoblastoid cells. *Proc. Natl. Acad. Sci. U.S.A.,* 81:3170–3174.
13. Dalla-Favera, R., Wong-Staal, F., and Gallo, R.C. (1982): One gene amplification in promyelocytic leukemia cell line HL-60 and primary leukemic cells of the same patient. *Nature,* 299:61–63.
14. Duesberg, P.H. (1985): Activated proto-onc genes: Sufficient or necessary for cancer? *Science,* 228:669–677.
15. Emanuel, B.S., Balaban, G., Boyd, J.P., Grossman, A., Negishi, M., Parmiter, A., and Glick, M.C. (1985): N-*myc* amplification in multiple homogeneously staining regions in two human neuroblastomas. *Proc. Natl. Acad. Sci. U.S.A.,* 82:3736–3740.
16. Fasano, O., Birnbaum, D., Edlund, L., Fogh, J., and Wigler, M. (1984): New human transforming genes detected by a tumorigenicity assay. *Mol. Cell. Biol.,* 4:1695–1705.
17. Feo, S., ar-Rushdi, A., Huebner, K., Finan, J., Nowell, C.P., Clarkson, B., and Croce, C.M. (1985): Suppression of the normal mouse c-*myc* oncogene in human lymphoma cells. *Nature,* 313:493–495.
18. Filmus J.E. and Buick, R.N. (1985): Stability of c-K-*ras* amplification during progression in a patient with adenocarcinoma of the ovary. *Cancer Res.,* 45:4468–4472.
19. Fujita, J., Srivastava, S.K., Krans, M.H., Rhim, J.S., Tronick, S.R., and Aaronson, S.A. (1985): Frequency of molecular alterations affecting *ras* proto oncogenes in human urinary tract tumors. *Proc. Natl. Acad. Sci. U.S.A.,* 82:3849–3853.
20. George, D.L. (1984): Amplification of cellular proto-oncogenes in tumours and tumor cell lines. *Cancer Surveys,* 3:497–513.
21. George, D.L., Scott, A.F., Martinville, B., and Francke, U. (1984): Amplified DNA in Y1 mouse adrenal tumor cells: isolation of cDNAs complementary to an amplified c-Ki-*ras* gene and localization of homologuous sequences to mouse chromosome 6. *Nucl. Acids Res.,* 12:2731–2743.
22. Graham, K.A., Richardson, C.L., Minden M.D., Trent, J.M., and Buick R.N. (1985): Varying degrees of amplification of the N-*ras* oncogene in the human breast cancer cell MCF-7. *Cancer Res.,* 45:2201–2205.
23. Hayashi, K., Fujiki, H., and Sugimura, T. (1983): Effects of tumor promoters on the frequency of methallothionein I gene amplification in cells exposed to cadmium. *Cancer Res.,* 43:5433–5436.
24. Hayashi, K., Kakizoe, T., and Sugimura T. (1983): *In vivo* amplification and rearrangement of the c-Ha-*ras*-1 sequence in a human bladder carcinoma. *Gann,* 74:798–801.
25. Hayashi, K., Makino, R., and Sugimura, T. (1984): Amplification and over-expression of the c-*myc* gene in Morris hepatomas. *Gann,* 75:475–478.

26. Heisterkamp, N., Stam, K., Groffen, J., de Klein, A., and Grosveld, G. (1985): Structural organization of the bcr gene and its role in the Ph' translocation. *Nature*, 315:758–761.
27. Hunter, T. (1984): Oncogenes and proto-oncogenes: How do they differ? *J. Natl. Cancer Inst.*, 73:773–786.
28. Kaczmarek, L., Hyland, K.J., Watt, R., Rosenberg, M., and Baserga, R. (1985): Microinjected c-myc as a competence factor. *Science*, 228:1313–1315.
29. Kelly, K., Cochran, B.H., Stiles, C.D., and Leder, P. (1983): Cell-specific regulation of the c-*myc* gene by lymphocyte mitogens and platelet-derived growth factor. *Cell*, 35:603–610.
30. Klein, G. (1981): The role of gene dosage and genetic transpositions in carcinogenesis. *Nature*, 294:313–318.
31. Klein, G., and Klein, E. (1984): Oncogene activation and tumor progression. *Caracinogenesis*, 5:429–435.
32. Klein, G., and Klein, E. (1985): Evolution of tumours and the impact of molecular oncology. *Nature*, 315:190–195.
33. Koda, T., Matsushima, S., Sasaki, A., Danjo, Y., and Kakinuma, M. (1985): c-*myc* gene amplification in primary stomach cancer. *Jpn. J. Cancer Res. (Gann)*, 76:551–554.
34. Kohl, N., Gee, C.E., and Alt, F.W. (1984): Activated expression of the N-*myc* gene in human neuroblastoma and related tumors. *Science*, 226:1335–1337.
35. Kohl, N.E., Kanda, N., Schreck, R.P., Bruns, G. Latt, S.A., Gilbert, F., and Alt, F.W. (1983): Transposition and amplification of oncogene-related sequences in human neuroblastomas. *Cell*. 35:359–367.
36. Konopka, J.B., Watanabe, S.M., and Witte, O.N. (1984): An alteration of the human c-*abl* protein in K562 leukemia cells unmasks associated tyrosine kinase activity. *Cell*, 37:1035–1042.
37. Kozbor, D., and Croce, C.M. (1984): Amplification of the c-*myc* oncogene in one of five human breast carcinoma cell lines. *Cancer Res.*, 44:438–441.
38. Land, H., Parada, L.F., and Weinberg, R.A. (1983): Tumorigenic conversion of primary embryo fibroblasts requires at least two cooperating oncogenes. *Nature*, 304:596–602.
39. Lee, W.H., Murphree, A.L., and Benedict, W.F. (1984): Expression and amplification of the N-*myc* gene in primary retinoblastoma. *Nature*, 309:458–460.
40. Libermann, T.A., Nusbaum, H.R., Razon, N., Kris, R., Lax, I., Soreq, H., Whittle, N., Waterfield, M.D., Ullrich, A., and Schlessinger, J. (1985): Amplification, enhanced expression and possible rearrangement of EGF receptor gene in primary human brain tumors of glial origin. *Nature*, 313:144–147.
41. Lin, C.R., Chen, W.S., Kruiger, W., Stolarsky, L.S., Weber, W., Evans, R.M., Verma, I.M., Gill, G.N., and Rosenfeld, M.G. (1984): Expression cloning of human EGF receptor complimentary DNA: gene amplification and three related messenger RNA products in A431 cells. *Science*, 224:845–848.
42. Little, C.D., Nau, M.M., Carney, D.N., Gazdar, A.F., and Minna, J. (1983): Amplification and expression of the c-*myc* oncogene in human lung cancer cell lines. *Nature*, 306:194–196.
43. Makino, R., Hayashi, K., and Sugimura, T. (1984): c-*myc* transcript is induced in rat liver at a very early stage of regeneration by cycloheximide treatment. *Nature*, 310:697–698.
44. Merlino, G.T., Xu, Y-H., Ishii, S., Clark, A.J.L., Semba, K., Toyoshima, K., Yamamoto, T., and Pastan, I. (1984): Amplification and enhanced expression of the epidermal growth factor receptor gene in A431 human carcinoma cells. *Science*, 224:417–419.
45. Miyaki, M., Sato, C., Matsui, T., Koike, M., Mori, T., Kosaki. G., Takai, S., Tonomura, A., and Tsuchida, N. (1985): Amplification and enhanced expression of cellular oncogene c-Ki-*ras*-2 in a human epidermoid carcinoma of the lung. *Jap. J. Cancer Res. (Gann)*, 76:260–265.
46. Moditahedi, N., Lavialle, C., Poupon, M.F., Landin, R.M., Cassingena, R., Monier, R., and Brison, O. (1985): Increased level of amplification of the c-*myc* oncogene in tumors induced in nude mice by a human breast carcinoma cell line. *Cancer Res.*, 45:4372–4379.
47. Moelling, K., Pfaff, E., Beug, H., Beimling, P., Buntc, T., Schaller, E.H., and Graf, T. (1985): DNA-binding activity is associated with purified *myb*-proteins from AMV and E26 viruses and is temperature-sensitive for E26 ts mutants. *Cell*, 40:983–990.
48. Montgomery, K.T., Biedler, J.L., Spengler, B.A., and Melera, P.W. (1983): Specific DNA sequence amplification in human neuroblastoma cells. *Proc. Natl. Acad. Sci. U.S.A.*, 80:5724–5728.
49. Murray, M.J., Cunningham, J.M., Parada, L.F., Dautry, F., Lebowitz, P., and Weinberg, R.A. (1983): The HL-60 transforming sequence: a *ras* oncogene coexisting with altered *myc* genes in hematopoietic tumors. *Cell*, 33:749–757.

50. Nakasato, F., Sakamoto, H., Mori, M., Hayashi, K., Shimosato, Y., Nishi, M., Takao, S., Nakatani, K., Terada, M., and Sugimura, T. (1984): Amplification of the c-*myc* oncogene in human stomach cancers. *Gann,* 75:737–742.
51. Nakatani, H., Tahara, E., Sakamoto, H., Terada, M., and Sugimura, T. (1985): Amplified DNA sequences in cancers. *Biochem. Biophys. Res. Commun.,* 130:508–514.
52. Nau, M.M., Carney, D.N., Battey, J., Johnson, B., Little, C., Gazdar, A., and Minna, J.D. (1984): Amplification, expression and rearrangement of c-*myc* and N-*myc* oncogenes in human lung cancer. *Curr. Top. Microbiol. Immunol.,* 113:172–177.
53. Newbold, R.F., and Overell, R.W. (1983): Fibroblast immortality is a prerequisite for transformation by EJ c-Ha-*ras* oncogene. *Nature,* 304:648–651.
54. Nishikura, K., Erikson, J., ar-Rushdi, A., Huebner, K., and Croce, C.M. (1985): The translocated c-*myc* oncogene of Raji Burkitt lymphoma cells is not expressed in human lymphoblastoid cells. *Proc. Natl. Acad. Sci. U.S.A.,* 82:2900–2904.
54a. Naguchi, S., Yokosuka, O., Tamanoi, F., Taya, Y., and Nishimura, S. (1986): *Submitted.*
55. Nowell, P., Finan, J., Dalla-Favera, R., Gallo, R.C., ArRushdi, A., Romanczuk, H., Selden, J.R., Emanuel, B.S., Rovera, G., and Croce, C.M. (1983): Association of amplified oncogene c-*myc* with an abnormally banded chromosome 8 in a human leukemia cell line. *Nature,* 306:494–497.
56. Pall, M.L. (1981): Gene-amplification model of carcinogeneesis. *Proc. Natl. Acad. Sci. U.S.A.,* 78:2465–2468.
57. Parada, L.F., Land, H., Weinberg, R.A., Wolf, D., and Rotter, V. (1984): Cooperation between gene encoding p53 tumour antigen and *ras* in cellular transformation. *Nature,* 312:649–651.
58. Pelicci, P.C., Lanfrancone, L., Brathwaite, M.D., Wolman, S.R., and Dalla-Favera, R. (1984): Amplification of the c-*myb* oncogene in a case of human acute myelogenous leukemia. *Science,* 224:1117–1121.
59. Persson, H., and Leder, P. (1984): Nuclear localization and DNA binding properties of a protein expressed by human c-*myc* oncogene. *Science,* 225:718–721.
60. Pfeifer-Ohlsson, S., Rydnert, J., Goustin, S.A., Larsson, E. Betsholts, C., and Ohlsson, R. (1985): Cell-type-specific pattern of *myc* protooncogene expression in developing human embryos. *Proc. Natl. Acad. Sci. U.S.A.,* 82:5050–5054.
61. Ralston, R., and Bishop, J.M. (1983): The protein products of the *myc* and *myb* oncogenes and adenovirus E1a are structurally related. *Nature,* 306:803–806.
62. Reitsma, P.H., Rothberg, P.G., Astrion, S.M., Trial, J., Bar-Shavit, Z., Teitelbaum, S.L., and Kahn, A.J. (1983): Regulation of *myc* gene expression in HL-60 leukaemia cells by a vitamin D metabolite. *Nature,* 306:492–494.
63. Roninson, I.B. (1983): Detection and mapping of homologous, repeated and amplified sequences by DNA renaturation in agarose gels. *Nucleic Acid Res.,* 11:5413–5431
64. Roninson, I.B., Abelson, H.T., Housman, D.E., Howell, N., and Varshavsky, A. (1984): Amplification of specific DNA sequences correlates with multi-drug resistance in Chinese hamster cells. *Nature* 309, 626–628.
65. Ruley, H.E. (1983): Adenovirus early region 1A enables viral and cellular transforming genes to transform primary cells in culture. *Nature,* 304:602–606.
66. Saksela, K., Bergh, J., Lehto, V-P., Nilsson, K., and Alitalo, K. (1985): Amplification of the c-*myc* oncogene in a subpopulation of human small cell lung cancer. *Cancer Res.,* 45:1823–1827.
67. Schimke, R.T. (1984): Gene amplification during resistance and cancer. *Cancer Res.,* 44:1735–1742.
68. Schwab, M., Alitalo, K., Klempnauer, K.H., Varmus, H.E., Bishop, J.M., Gilbert, F., Brodeur, G., Goldstein, M., and Trent, J. (1983): Amplified DNA with limited homology to *myc* cellular oncogene is shared by human neuroblastoma cell lines and a neuroblastoma tumor. *Nature,* 305:245–248.
69. Schwab, M., Alitalo, K., Varmus, H.E., Bishop, J.M., and George, D. (1983): A cellular oncogene (c-Ki-*ras*) is amplified, overexpressed, and located within karyotype abnormalities in mouse adrenocortical tumor cells. *Nature,* 303:497–501.
70. Schwab, M., Ellison, J., Busch, M., Rosenau, W., Varmus, H.E., and Bishop, J.M. (1984): Enhanced expression of the human gene N-*myc* consequent to amplification of DNA may contribute to malignant progression of neuroblastoma. *Proc. Natl. Acad. Sci. U.S.A.,* 81:4940–4944.
71. Schwab, M., Ramsay, G., Alitalo, K., Varmus, H.E., Bishop, J.M., Martinsson, T., Levan G., and Levan, A. (1985): Amplification and enhanced expression of the c-*myc* oncogene in mouse SEWA tumour cells. *Nature,* 315:345–347.

72. Schwab, M., Varmus, H.E., and Bishop, J.M. (1985): Human N-*myc* gene contributes to neoplastic transformation of mammalian cells in culture. *Nature,* 316:160–162.
73. Schwab, M., Varmus, H.E., Bishop, J.M., Grzeschik, K.H., Naylor, S.L., Sakguchi, A.Y., Brodeur, G., and Trent, J. (1984): Chromosomal localization in normal human cells and neuroblastomas of a gene related to c-*myc. Nature,* 308:288–291.
74. Seki, T., Fujii, G., Mori, S., Tamaoki, N., and Shibuya, M. (1985): Amplification of c-*yes*-1 proto-oncogene in a primary stomach cancer. *Jpn. J. Cancer Res. (Gann),* 76:907–910.
75. Sekiya, T., Fushimi, M., Hirohashi, S., and Tokunaga, A. (1985): Amplification of activated c-Ha-*ras*-1 in human melanoma. *Jap. J. Cancer Res. (Gann),* 76:555–558.
76. Selden, J.R., Emanuel, B.S., Wang, E., Cannizzaro, L., Palumbo, A., Erikson, J., Nowell, P.C., Rovera, G., and Croce, C.M. (1983): Amplified Cλ and c-*abl* genes are on the same marker chromosome in K562 leukemia cells. *Proc. Natl. Acad. Sci. U.S.A.,* 80:7289–7292.
77. Shibuya, M., Yokota, J., and Ueyama, Y. (1985): Amplification and expression of a cellular oncogene (c-*myc*) in human gastric adenocarcinoma cells. *Mol. Cell. Biol.,* 5:414–418.
78. Shtivelman, E., Lifshits, B., Gale, P.R., and Canaani, E. (1985): Fused transcript of *abl* and *bcr* genes in chronic myelogenous leukaemia. *Nature,* 315:550–554.
79. Spandidos, D.A., and Wilkie, N.M. (1984): Malignant transformation of early passage rodent cells by a single mutated human oncogene. *Nature,* 310:469–475.
80. Stark, G.R., and Wahl, G.M. (1984): Gene amplification. *Ann. Rev. Biochem.,* 53:447–491.
81. Taya, Y., Hosogai, K., Hirohashi, S., Shimosato, Y., Tsuchiya, R., Tsuchida, N., Fushyimi, M., Sekiya, T., and Nishimura, S. (1984): A novel combination of K-*ras* and *myc* amplification accompanied by point mutational activation of K-*ras* in a human lung cancer. *EMBO J.,* 3:2943–2946.
82. Thiele, C.J., Reynolds, C.P., and Israel, M.A. (1985): Decreased expression of N-*myc* precedes retinoic acid-induced morphological differentiation of human neuroblastoma. *Nature,* 313:404–406.
83. Ullrich, A., Coussens, L., Hayaflick, J.S., Dull, T.J., Gray A., Tam, A.W., Lee, J., Yarden, Y., Libermann, T., Schlessinger, J., Downward, J., Mayes, E.L.V., Whittle, Waterfield, M.D., and Seeburg, P.H. (1984): Human epidermal growth factor receptor cDNA sequence and aberrant expression of the amplified gene in A431 epidermoid carcinoma cells. *Nature,* 309:418–425.
84. Varshavsky, A. (1981): On the possibility of metabolic control of replicon "misfiring": relationship to emergence of malignant phenotypes in mammalian cell lineages. *Proc. Natl. Acad. Sci. U.S.A.,* 78:3673–3677.
85. Westin, E.H., Gallo, R.C., Arya, S.K., Eva, A., Souza, L.M., Baluda, M.A., Aaronson, S.A., and Wong-Staal, F. (1982): Differential expression of the *amv* gene in human hematopoietic cells. *Proc. Natl. Acad. Sci. U.S.A.,* 79:2194–2198.
85a. Yamada, H., Sakamoto, M., Taira, Y., Nishimura, S., Shimosato, M., Terada, M., and Sugimura, T. (1986): *Jpn. J. Cancer Res. (Gann),* 77:370–375.
86. Yamamoto, T., Kamata, N., Kawano, H., Shimizu, S., Kuroki, T., Toyoshima, K., Rikimaru, K., Nomura, N., Ishizaki, R., Pastan, I., Gamou, S., and Shimizu, N. (1986): High incidence of amplification of the EGF recept gene in human squamous carcinoma cell lines. *Cancer Res.,* 46:414–416.
87. Yancopoulos, G.D., Nisen, P.D., Tesfaye, A., Kohl, N.E., Godfarb, M.P., and Alt, F.W. (1985): N-*myc* can cooperate with *ras* to transform normal cells in culture. *Proc. Natl. Acad. Sci. U.S.A.,* 82:5455–5459.
88. Yander, G., Halsey, H., Kenna, M., and Augenlicht, L.H. (1985): Amplification and elevated expression of c-*myc* in a chemically induced mouse colon tumor. *Cancer Res.,* 45:4433–4438.

Amplification and Expression of the *myc* Gene in Small-Cell Lung Cancer

Burke J. Brooks, Jr., James Battey, Marion M. Nau, Adi F. Gazdar, and John D. Minna

NCI-Navy Medical Oncology Branch, National Cancer Institute National Institutes of Health and Naval Hospital, and the Department of Medicine, Uniformed Service University of the Health Sciences, Bethesda, Maryland 20814

Lung cancer is the most common fatal malignancy in the United States, and its frequency is rising rapidly around the world. In 1985 in the United States, it was estimated that there would be 144,000 new cases of lung cancer with 125,000 deaths, making this malignancy a major source of morbidity and mortality (28). To gain new methods for preventing and treating this disease, our laboratory has studied the biology of lung cancer and has established a large number of long-term continuous cell cultures of human lung cancer (3,7). These cell lines have greatly facilitated the study of this tumor.

Human lung cancer can be divided pathologically and biologically into small-cell lung cancer (SCLC), representing 25% of cases, and non-small-cell lung cancer (NSCLC) (squamous, adenocarcinoma, and large-cell cancer), representing 75% of cases. In 1983, we reported our observation that the proto-oncogene, c-*myc*, was amplified and expressed in certain SCLC cell lines (14). Subsequently, we have found two other *myc*-related genes (N-*myc* and L-*myc*) amplified and/or expressed in SCLC (17,18). Thus, we have begun to define both structurally and functionally this *myc* gene family and its relationship to the biology of SCLC.

CELL CULTURE AND BIOCHEMICAL CHARACTERIZATION

Since 1975, our laboratory has developed methods to establish SCLC in long-term continuous culture. To date, we have established approximately 60 SCLC cell lines from patient tumor specimens and nine cell lines from nude mouse xenografts of transplanted human tumors. Along with serum-containing media, we have also used a serum-free defined media, which allows growth of some tumor lines unable to grow in serum-supplemented media (2,29). SCLC cell lines usually grow in tightly packed spherical aggregates, have a characteristic cytologic and

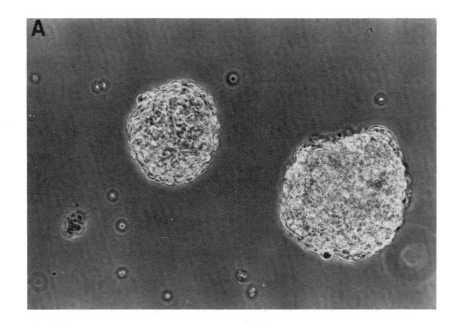

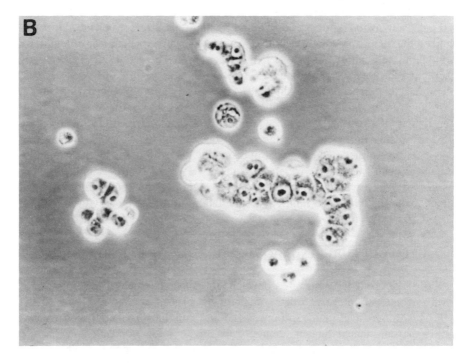

Fig. 1. *In vitro phase*-contrast morphological characteristics of SCLC cell lines in culture. **A:** Classic cell line. **B:** Variant cell line.

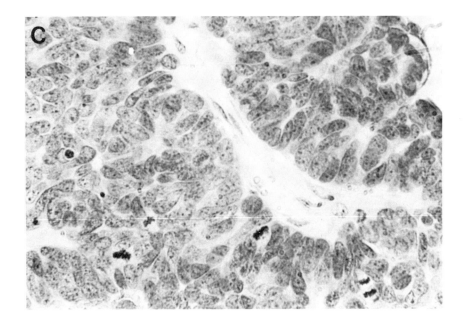

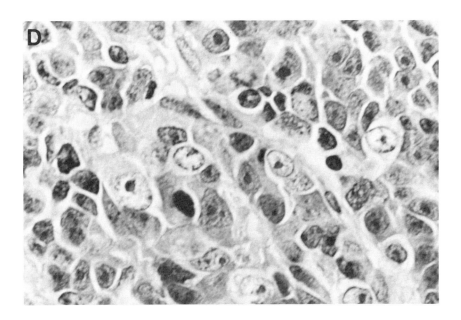

Fig. 1. (cont.) **C:** Histologic appearance of classic cell line. **D:** Histologic appearance of variant cell line. Hematoxylin and eosin stain 100× magnification. (See text for description.)

TABLE 1. Properties of classic and variant SCLC cell lines[a]

Property	Classical	Variant
Morphology	Tightly packed spheres	Loosely arranged
Colony-forming efficiency (%)	2	14
HNK-1 (Leu-7) antigen	Present	Present
Dense core granules	Present	Absent
DDC (units/mg)[b]	199 ± 32.9	<1
BLI (pmol/mg)[b]	3.97 ± 0.90	<0.01
NSE (ng/mg)[b]	1495 ± 247	481 ± 87
CK-BB (ng/mg)[c]	6488 ± 829	5936 ± 990
c-myc Amplification	1/23[d]	7/9

[a](DDC) dopa decarboxylase; (BLI) bombesin immunoreactivity; (NSE) neuron-specific enolase; (CK-BB) creatine kinase-BB; all units ± S.E. From refs. 3 and 6.
[b]$p < 0.01$.
[c]$p = 0.94$.
[d]Line N390 has borderline c-myc DNA amplification (<4-fold) and does not express c-myc mRNA.

histologic appearance, and express the entire spectrum of SCLC neuroendocrine biochemistry (Fig. 1a,c; Table 1). They are referred to as classic SCLC (3). A subset of SCLC cell lines (30%), which has the appearance in culture of free-floating, loosely aggregated cells, is called variant SCLC cell lines (Fig. 1b) (3,6). Histologically, variant cells have abundant cytoplasm and large, pale nuclei with distinct features not found in classic SCLC (Fig. 1d). Both variant and classic SCLC cell lines show morphologic and biochemical properties of neuroendocrine cells, such as the BB isozyme of creatine phosphokinase (CK-BB), neuronspecific enolase (NSE), and the neuroendocrine antigen marker HNK-1 (Leu-7) (1,6,7). Compared to classic SCLC, variant cell lines have a shorter doubling time, higher cloning efficiency, increased tumorgenecity in athymic nude mice, are more resistant to radiation *in vitro,* and do not express all the neuroendocrine markers that classic SCLC cell lines do. Variant cell lines fail to produce dense core (neurosecretory) granules, high levels of the enzymes dopa decarboxylase (DDC), and the bombesin-related peptide (gastrin-releasing peptide) that are usually found in classic SCLC (Table 1) (6).

Cytogenetic analysis of some SCLC cell lines have revealed the presence of double-minute chromosomes (DMs) or chromosomes with homogeneously staining regions (HSRs) (31,32). Because of these cytogenetic abnormalities and their association with gene amplification (4,22,23), we speculated that these cell lines might reveal amplified gene sequences that might be involved in the biological behavior of SCLC, either related to growth or drug resistance. With this in mind, we began screening our SCLC cell lines containing HSRs and DMs with oncogene- and drug-resistance gene probes.

c-*myc* GENE AMPLIFICATION AND EXPRESSION IN SCLC AND THE VARIANT SUBTYPE

In an initial study of 17 cell lines from patients with SCLC (14), we observed that several SCLC cell lines of the variant subtype were amplified (Fig. 2a) and also showed increased levels of c-*myc* mRNA expression (Fig. 2b), using the 1.6 kilobase (kb) *Sst*-I second exon c-*myc* probe. We have found amplification and/or expression of the c-*myc* oncogene in seven of nine variant cell lines. However, one variant cell line (H526) not amplified for c-*myc* has DMs and exhibits amplification of another oncogene, N-*myc* (see later in text), whereas the other (H433) expresses another *myc*-related gene, L-*myc* (data not shown). One classic cell line (H146) has increased steady-state expression of c-*myc* mRNA but no detectable DNA amplification. Thus, it appears that amplification and expression of the c-*myc* gene are associated with alterations in growth and morphological characteristics of the variant subset of SCLC cell lines (6).

We have examined more than 20 tumor specimens from autopsies of patients, who at diagnosis had classic SCLC for the presence of c-*myc* amplification. Although we have been able to readily identify amplification and/or expression of the c-*myc* gene in cell lines established from patients with SCLC, we have been unable to demonstrate amplification of the c-*myc* gene in direct tumor specimens from patients. Why we have been able to establish variant cell lines with c-*myc* amplification remains unclear; however, one reasonable hypothesis is that cell culture provides a selective advantage for the outgrowth of a small number of tumor cells in the biopsy specimen containing amplified c-*myc* genes. For example, cell line, N231, a classical SCLC cell line has fivefold c-*myc* amplification. During a 1-year period of culture, a rapidly growing line N417 of the variant morphology with 40-fold c-*myc* DNA amplification was derived from the nude mouse xenograft of this line.

N-*myc* GENE AMPLIFICATION AND EXPRESSION IN SCLC

While screening for c-*myc* amplification in SCLC, we observed that several SCLC lines contained a 2.0-kb *Eco* RI fragment that hybridized to a human second exon c-*myc* probe (fig. 3a) (18). We were attempting to clone this fragment when two laboratories reported the isolation and partial characterization of a similar 2.0-kb *Eco* RI fragment. This DNA was found to be amplified in human neuroblastomas and subsequently in retinoblastomas, both tumors with neuroendocrine properties (10,13,24). This gene (called N-*myc* after its initial isolation from a neuroblastoma cell line) was found to have limited areas with distinct DNA sequence homology to sections of the second and third exons of the c-*myc* gene (10,11,13,15,18). Mapping studies assigned the N-*myc* gene to human chromosome 2*p* (26), in contrast to the c-*myc* proto-oncogene located on chromosome 8*p* (30).

Because of the multiple neuroendocrine properties in common between SCLC and neuroblastoma, it seemed likely that the new 2.0-kb *Eco* RI fragment we detected was related to the N-*myc* gene. In fact, using a human N-*myc* probe (Nb-1) (24), we identified five SCLC cell lines with an amplified 2.0-kb *Eco* RI N-*myc* fragment. In addition, one cell line, H526, exhibits an additional 5.5-kb *Eco* RI genomic fragment (detected by the N-*myc* probe) that is amplified (Fig. 3b). The exact relationship of this novel 5.5-kb fragment to the 2.0-kb *Eco* RI

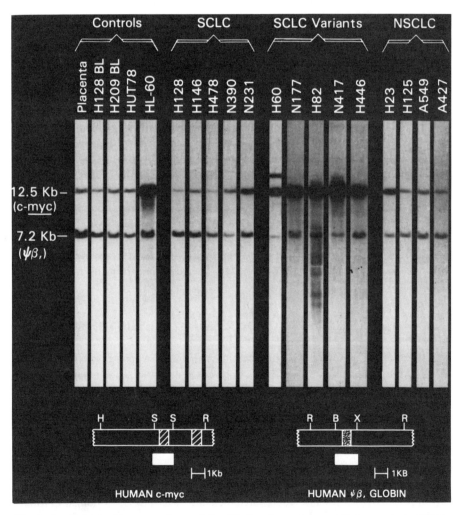

Fig. 2. A: Hybridization comparisons of *Eco* RI digestions of representative controls, SCLC, SCLC variants, and non-SCLC cell line DNAs with second exon human c-*myc* and human pusdeo β₁ (ψβ₁) globin probes (5): (B) *Bgl* II; (E) *Eco* RI; (HY) *Hind* III; (S) *Sst* I; (X) *Xba* I.

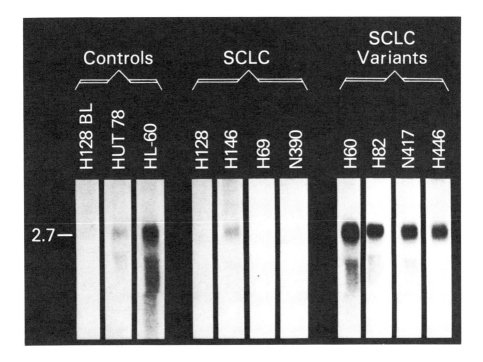

Fig. 2. (cont.) B: Hybridization of control, SCLC, and SCLC variant cell line total RNAs to the second exon human c-*myc* gene. (For methods, see ref. 14.)

fragment has not yet been determined. It might be the result of a N-*myc* somatic DNA mutation, rearrangement of the N-*myc* gene during amplification, or a new N-*myc*-related gene is this cell line. It is interesting to note that just as the second exon c-*myc* probe detects N-*myc*-related sequences in lines amplified for N-*myc* (fig. 3a, H249, H526, H689), the N-*myc* probe, containing a homologous sequence to the second exon the of c-*myc* gene, can detect the c-*myc* sequence in cell lines amplified for this gene (Fig. 3b, H446). Thus, this same approach would have allowed the discovery of the c-*myc* gene using an N-*myc* probe if the N-*myc* gene had been discovered before the c-*myc* gene. This concept had important implications for our subsequent work.

A 3.1-kb N-*myc* hybridizing transcript is identified in the poly A (+) RNA of each of the five cell lines with N-*myc* DNA amplification, as well as in poly A (+) RNA from a neuroblastoma cell line (KCNR) (Fig. 4a). One cell line, H187, which was not amplified for the N-*myc* gene, also expresses the 3.1-kb transcript (Fig. 4). Cell line H526, which has both the 2.0- and 5.5-kb *Eco* RI N-*myc* hybridizing fragments, expresses a novel 1.8-kb transcript in addition to the 3.1-kb transcript. A transcript of approximately the same size has been seen in neuroblas-

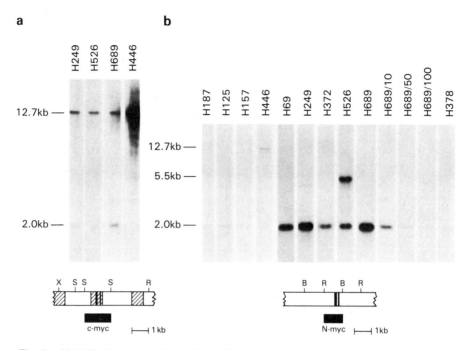

Fig. 3. Hybridization comparisons of Eco RI digestions of SCLC and non-SCLC cell line DNAs with (**a**) human second exon c-*myc* and (**b**) N-*myc* (Nb-1) probes. The 12.7-kb position represents the Eco RI germline c-*myc* gene fragment,. and the 2.0-kb position is the Eco RI germ-line N-*myc* fragment. (See ref. 18 for methods.) In b, 689/10, 689/50, and 689/100 represent 1:10, 1:50, 1:100 dilutions, respectively, of H689 DNA. In this and subsequent schematic probe diagrams, *bold vertical lines* indicate regions of homology between c-, N-, and L-*myc* DNA sequences, (B) Bam HI; (R) Eco RI; (S) Sst I; (X) Xho I.

toma cell lines that are amplified and express the N-*myc* gene (15). Whether or not this 1.8-kb transcript originates from the amplified 5.5-kb Eco RI fragment remains to be studied.

In order to study the structure of the N-*myc* gene in greater detail, we cloned approximately 12 kb of genomic DNA surrounding the 2.0-kb Eco RI N-*myc* fragment from a human placental library (20). An additional N-*myc* exon with homology to the third exon of the c-*myc* gene was identified in this genomic clone on a 2.2-kb Bgl II–Eco RI fragment. As expected, this fragment also hybridizes to the 3.1-kb N-*myc* mRNA (Fig. 4b), which is in agreement with other reports linking this exon-containing fragment of the N-*myc* gene to the third exon of c-*myc* (11,15,25).

In contrast to the c-*myc* gene where we have not yet been able to show amplification in actual tumor specimens, we have two examples of N-*myc* amplification in tumor specimens taken directly from previously treated patients (Fig, 5b,c;

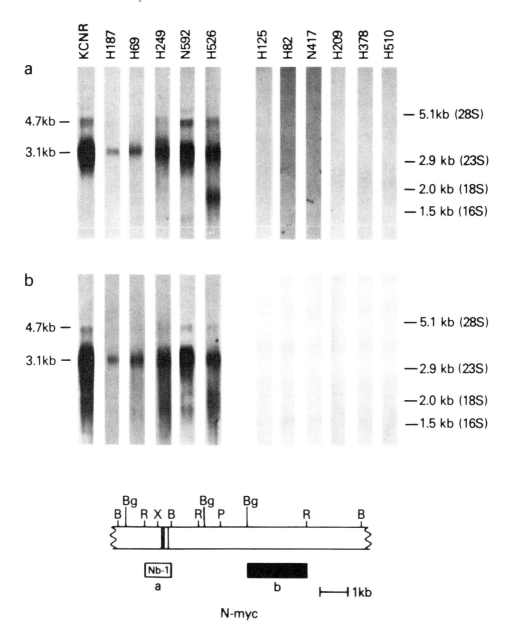

Fig. 4. Hybridization of poly (A)+ cell line RNAs to (a) the 1.0-kb Eco RI–Bam-HI N-myc probe (Nb-1) and (b) the 2.2-kb Bgl II–Eco RI 3' placental N-myc probe. The six lanes on the right side of both a and b (all c- or L-myc expressing cell lines) show no detectable N-myc hybridization. (See ref. 18 for methods.)

H526, H689), whereas noninvolved tissue does not reveal amplification. Tumor specimen H720 was taken from a previously untreated patient (Fig. 5d). It, too, revealed N-*myc* gene amplification, thus indicating amplification can occur prior to treatment. It is interesting to note that the 5.5-kb *Eco* RI-amplified fragment found in the cell line H526, established from a bone marrow aspiration, was present in only one of three liver metastases at autopsy (Fig. 5b). All tumor tissue contained the 2.0-kb *Eco* RI-amplified N-*myc* fragment. Clearly, these findings indicate a heterogeneity of tumor cells carrying N-*myc* gene amplification fragments in this patient.

Loss of amplified sequences [e.g., the dihydrofolate reductase (*dhfr*) gene] can occur when selection pressure is released (22,23). We have reported that SCLC tumor line H249 lost amplified *dhfr* gene sequences when grown in cell culture in the absense of methotrexate (4). However, no apparent change in the amount of N-*myc* gene amplification was noted in this cell line (H249) and in cell line H526 (Fig. 5a) after continuous growth in culture for >30 and >15 weeks (70 and 50 population doublings), respectively.

When the tumor tissue of H720 was placed in culture, there was evidence that amplification increased during the initial passages (Fig. 6). Additional novel, hybridizing N-*myc Eco* RI fragments of 5.5 and 4.8 kb appeared during serial culture. The increase in amplification seen in culture may result from a selective growth advantage for cells producing the N-*myc* gene product. As with H526, the structural and functional meaning of these additional N-*myc* hybridizing fragments are not known at present. They could represent rearrangement of the N-*myc* gene occurring with amplification or amplification of an N-*myc*-related gene. In our

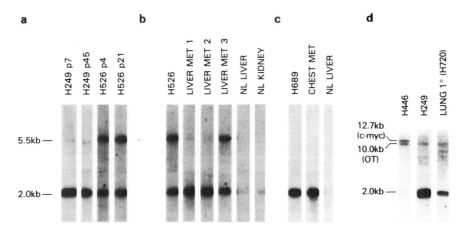

Fig. 5. Hybridization of *Eco* RI digestions of SCLC cell line DNAs (**a**) (H249, passages 7 and 45, and H526, passage 4 and 21); (**b**) patient H526; (**c**) patient H689 DNAs from cell line, as well as tumor and normal tissue samples harvested directly from the patient at autopsy; (**d**) SCLC DNAs (*c-myc* amplified cell line, H446; N-*myc* amplified cell line, H249) and primary tumor (H720) to 1.0-kb *Eco* RI–*Bam* HI (Nb-1) probe; OT represents a human oxytocin probe as a reference single copy gene (21). (See ref. 18 for methods.)

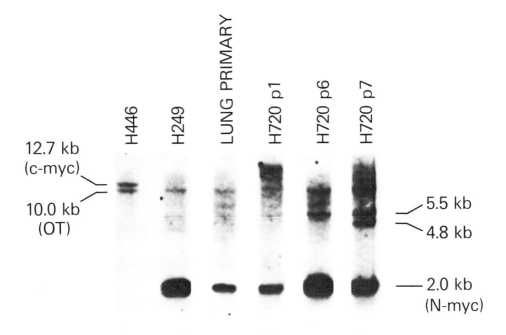

Fig. 6. Hybridization of Eco RI digestions of primary tumor DNA and subsequent passage in cell culture DNAs, along with SCLC DNAs (c-*myc* amplified cell line, H446; N-*myc* amplified cell line H249) to 1.0-kb *Eco* RI–*Bam* HI (Nb-1) probe; OT represents a human oxytocin probe used as a reference single-copy gene (21).

work with cell line H720, we demonstrate that significant changes in amplification occur in the first few passages *in vitro*.

L-*myc* GENE AMPLIFICATION AND EXPRESSION IN SCLC

During our studies of SCLC genomic DNA using an N-*myc* probe (Nb-1), which has a homology to the second exon of the c-*myc* gene, we occasionally detected a 10.0- or 6.6-kb *Eco* RI genomic fragment in DNA samples that were not amplified for either the c- or N-*myc* genes (Fig. 7) (17). To understand better the structure of these *myc*-related species, the 10.0-kb fragment was cloned from *Eco* RI-digested, size-fractionated SCLC cell line H378 genomic DNA. Restriction endonuclease mapping and DNA sequence analysis of portions of this clone revealed that this cloned fragment had two small regions of homology but was otherwise unrelated in sequence to either the c-*myc* or N-*myc* genes. We termed the new *myc*-related sequence L-*myc*, since it was first isolated from a lung tumor cell line. A 1.8-kb *Sma* I–*Eco* RI subfragment of the clone located about 100 base pairs

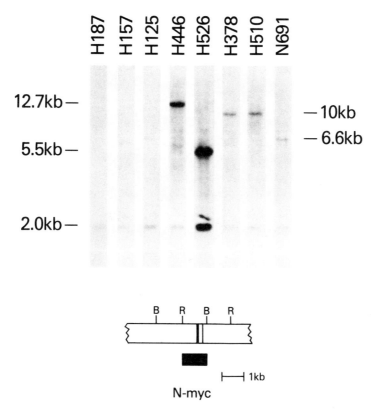

Fig. 7. Detection of novel *myc*-related sequences in *Eco* RI digestions of SCLC cell line DNAs with a human N-*myc* probe Nb-1. The 12.7-kb position represents the *Eco* RI germline c-*myc* gene fragment; the 2.0-kb position, the *Eco* RI germline N-*myc* fragment; the 5.5-kb position, a putative N-*myc*-related gene sequence; the 10.0- and 6.6-kb positions, the *Eco* RI *myc*-related gene fragments representing L-*myc*; (B) *Bam* HI; (R) *Eco* RI; (S) *Sma* I; (Bg *Bgl* II. (See ref. 17 for methods.)

(bp) 3' to sequences homologous to the c-*myc* and N-*myc* genes in the L-*myc* clone (see restriction map, fig. 8) was subcloned. This fragment (*Sma* I–*Eco* RI) did not cross hybridize with either c- or N-*myc* sequences. Using this 1.8-kb *Sma* I–*Eco* RI fragment as a probe, we examined *Eco* RI digests of SCLC DNAs (Fig. 8) as well as normal and other tumor DNAs in Southern blot experiments (data not shown). In addition to the expected 10.0-kb *Eco* RI DNA fragment, we also observed hybridization to a 6.6-kb *Eco* RI fragment in several DNAs. Both normal and tumor DNA samples were identified that contain both *Eco* RI fragments or only one of the two fragments (10.0 and 6.6 kb). Since other restriction endonuclease digests show no pattern differences in these same DNA samples, these two *Eco* RI fragments represent an *Eco* RI restriction-fragment-length polymorphism (RFLP) for the L-*myc* gene. The same two alleles (10.0 and 6.6 kb) are observed

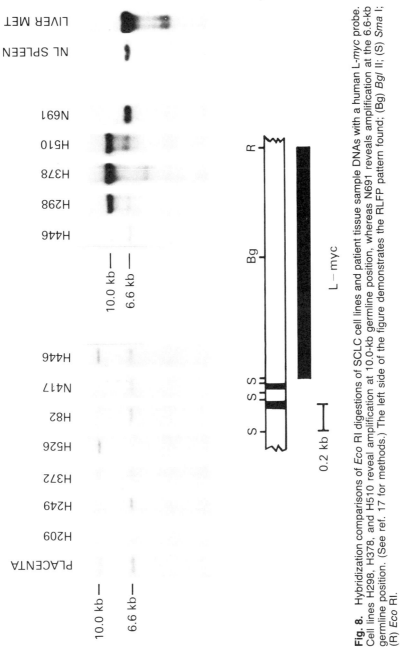

Fig. 8. Hybridization comparisons of *Eco* RI digestions of SCLC cell lines and patient tissue sample DNAs with a human L-*myc* probe. Cell lines H298, H378, and H510 reveal amplification at 10.0-kb germline position, whereas N691 reveals amplification at the 6.6-kb germline position. (See ref. 17 for methods.) The left side of the figure demonstrates the RLFP pattern found; (Bg) *Bgl* II; (S) *Sma* I; (R) *Eco* RI.

with an equal frequency distribution in both SCLC and other tumor DNAs and lymphocyte DNA from normal individuals as well, indicating that these two alleles represent a RFLP rather than a tumor-specific rearrangement. In addition, these data show to date that neither allele (10.0 or 6.6 kb) is specifically associated with SCLC tumorigenesis.

Hybridization studies using the 1.8-kb *Sma I–Eco* RI L-*myc*-specific probe revealed the DNAs to be amplified for either the 10.0- or the 6.6-kb *Eco* RI fragment (Fig. 8). In DNAs containing both L-*myc* alleles (10.0 and 6.6 kb), only one of the two alleles was found to be amplified. In addition, a SCLC tumor sample obtained directly from an autopsy tissue specimen shows about a 15-fold DNA amplification of the L-*myc* gene when compared to the patient's normal tissue DNA. Similarly, just as detection of the novel fragments in human neuroblastoma DNAs using the c-*myc* or v-*myc* probes led to the discovery of amplified N-*myc* genes, the novel fragments found in lung cancer cell line DNAs using an N-*myc* probe led to the discovery of amplified L-*myc* fragments.

L-*myc* gene expression was examined in a group of SCLS lines, including all cell lines demonstrating L-*myc* DNA amplification (Fig. 9). Cell lines amplified

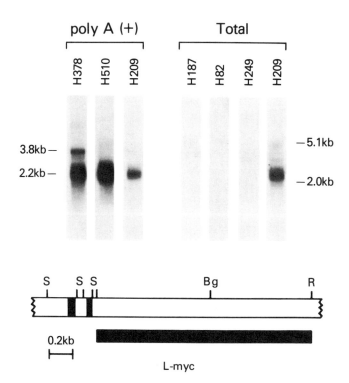

Fig. 9. Hybridization comparisons of poly (A)+ and total SCLC cell line RNAs with a human L-*myc* probe. The 5.1- and 2.0-kb positions indicate 28 S and 18 S human ribosomal RNAs respectively: (Bg) *Bgl* II; (S) *Sma* I; (R) *Eco* RI. (See ref. 17 for methods.)

for L-*myc* DNA sequences show a 2.2-kb poly A (+) transcript hybridizing to the L-*myc*-specific probe (1.8 kb *Sma* I–*Eco* RI). Poly A(+) RNA from H378 also shows a prominent 3.8-kb RNA transcript. One other SCLC cell line (H209) not amplified for L-*myc* DNA also expresses an L-*myc* specific transcript. The relationship between the 2.2- and 3.8-kb RNA transcripts is currently under investigation.

Previous experiments comparing c-*myc* and N-*myc* sequences have demonstrated two short regions of homology between these two genes, localized to the second exon of the C-*myc* gene (24). To better understand the structural relationship of the L-*myc* gene to the c- and N-*myc* genes, the region of homology in L-*myc* (designated by heavy vertical lines in the figures) was mapped to restriction fragments that are located about 100 bp 5' to the 1.8-kb *Sma* I–*Eco* RI L-*myc* fragment and the nucleotide sequence determined (Fig. 10). Like N-*myc*, L-*myc* also contains the same two regions of homology to the second exon of the c-*myc* gene separated by a stretch of nucleotides, which bears no resemblance to the analagous region in c-*myc* or N-*myc*. An open reading frame spans the 5' and 3' blocks of homology shown in Fig. 10, including the nonhomologous stretch between these two regions in all the genes. Both the nucleotide and predicted amino acid sequence show an 80% homology among the c-, N-, and L-*myc* genes.

The C-*myc* gene is assigned to human chromosome 8 (30), whereas N-*myc* is assigned to chromosome 2 (26) indicating dispersion of these two genes to different autosomes. We mapped the L-*myc* gene to a humam chromosomal region using both somatic cell hybrid analysis and human chromosome *in situ* hybridization. Both analyses localized the L-*myc* gene to human chromosome 1p with the

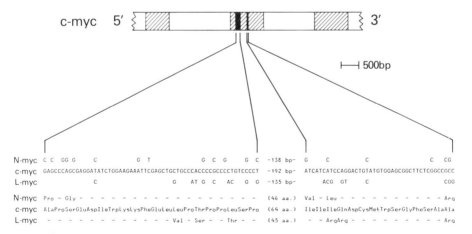

Fig. 10. Nucleotide and predicted amino acid sequence comparison of c-, N-, and L-*myc* homologous regions. At both the nucleotide and amino acid levels, the c-*myc* homology region sequence is shown, whereas only *differences* from c-*myc* are indicated in the N- and L-*myc* sequences. The numbers between the homology regions represent nucleotides and amino acids separating these regions respectively. (See ref. 17 for methods.)

in situ localization at or near band 1*p*32 (17). This cytogenetic localization is clearly distinct from both the c- and N-*myc* genes and is approximately the same location to which an earlier oncogene, B-*lym* (16), was mapped, but distal to the proto-oncogene N-*ras* (19). It is interesting to note that the human melanoma kindred gene has also been provisionally localized to 1*p* (8).

BIOLOGY OF THE *myc* GENE IN SCLC

In a study of neuroblastoma, the presence of N-*myc* amplification has been correlated with increased clinical stage and more rapid tumor progression (27). What is the significance of finding a *myc*-related gene amplified in approximately 60% of the cell lines studied? Does the presence of either amplification and/or expression of a *myc*-related gene further increase the already malignant behavior of SCLC? We have shown that when the c-*myc* gene is either amplified and/or expressed in cell lines, the variant characteristics (Table 1) are found almost exclusively. There has been no such correlation found with either N-*myc* or L-*myc* involvement in SCLC. In a retrospective review of all patients who had a cell line established during their clinical course, we found that amplification of either the c-*myc* or N-*myc* gene was an independent factor in a shorter survival time compared with patients who did not have cell lines amplified for one of these two genes (9). We are currently investigating this observation further in a prospective manner in untreated patients, since most of our work has involved previously treated and relapsed patients.

Both specific constructions of c-*myc* and N-*myc* have been shown to complement the *ras* proto-oncogene, forming foci in primary rat embryo fibroblasts after DNA transfection (12,25,33). Can L-*myc* also complement the *ras* oncogene in transforming primary fibroblasts after DNA transfection and is a *myc* gene important in the tumorigenesis of SCLC? Both of these questions are currently being addressed in our laboratory.

Are there differences in *myc* gene involvement between SCLC and NSCLC? How heterogeneous is the expression of *myc* genes between patients and/or within an individual tumor sample? Hopefully, both single-cell *in situ* hybridization and immunohistochemical assays on direct tumor specimens with various *myc* gene product probes will help to answer such questions. Are there other *myc*-related genes activated in the remaining SCLC cell lines without c-, N-, or L-*myc* amplification and/or expression? We have detected additional *myc*-related fragments in genomic DNA and are currently involved in their characterization.

CONCLUSIONS

We have found that one member of the *myc* gene family is either amplified and/or expressed in approximately 60% of our SCLC cell lines. C-*myc* amplification appears to be correlated with the more malignant behavior of the variant cell histology of SCLC, whereas the role of N-*myc* and L-*myc* in the biology of SCLC remains to be defined. The highly conserved region among the three *myc* genes suggests that it may play an important role in the function of the *myc* gene protein products. Studies currently underway into the clinical and biological significance of these genes in SCLC should add new understanding to this disease.

ACKNOWLEDGMENTS

We thank Philip Leder for his human c-*myc* clones and human placenta library; J. Michael Bishop and Manfred Schwab for the N-*myc* clone; Patrick Reynolds for cell line SMS-KCNR; Sylvia Stephenson for cell culture; and Tanya Myers for typing of the manuscript. The opinions and assertions contained herein are the private views of the authors and are not to be construed as official or as reflecting the views of the Department of the Navy or the Department of Defense.

REFERENCES

1. Bunn, P.A., Linnoila, I., Minna, J.D., Carney, D., and Gazdar, A.F. (1985): Small cell lung cancer, endocrine cells of the fetal bronchus, and other neuroendocrine cells express the Leu-7 antigenic determinant present on natural killer cells. *Blood,* 65:764–768.
2. Carney, D.N., Bunn, P.A., Gazdar, A.F., Pagan, J.F., and Minna, J. (1981): Selective growth in serum-free hormone-supplemented medium of tumor cells obtained by biopsy from patients with small cell carcinoma of the lung. *Proc. Natl. Acad. Sci. U.S.A.*, 78:3185–3189.
3. Carney, D.N., Gazdar, A.F., Bepler, G., Guccion, J.G., Marangos, P.J., Moody, T.W., Zweig, M.H., and Minna, J.D. (1985): Establishment and identification of small cell lung cancer cell lines having classic and variant features. *Cancer Res.*, 45:2913–2923.
4. Curt, G.A., Carney, D.N., Cowan, K.H., Jolivet, J., Bailey, B.D., Drake, J.C., Kao-Shan, C.S., Minna, J.D., and Chabner, B.A. (1983): Unstable methotrexate resistance in human small-cell carcinoma associated with double minute chromosome. *N. Eng. J. Med.*, 208:199–202.
5. Fritisch, E., Lawn, R.M., and Maniatis, T. (1980): Molecular cloning and characterization of the human B-like globulin gene cluster. *Cell,* 19:959–972.
6. Gazdar, A.F., Carney, D.N., Nau, M.M., and Minna, J.D. (1985): Characterization of variant subclasses of cell lines derived from small cell lung cancer having distinctive biochemical, morphological, and growth properties. *Cancer Res.*, 45:2924–2930.
7. Gazdar, A.F., Zweig, M.H., Carney, D.N., Van Steirteghen, A.C., Baylin, S.B., and Minna, J.D. (1981): Levels of creatine kinase and its BB isoenzyme in lung cancer specimens and cultures. *Cancer Res.*, 41:2773–2777.
8. Greene, M.H., Goldin, L.R., Clark, W.H., Jr., Lovrien, E., Kraemer, K.H., Tucker, M.A., Elder, D.E., Fraser, M.C., and Rowe, S. (1983): Familial cutaneous malignant melanoma: Autosomal dominant trait possibly linked to the Rh locus. *Proc. Natl. Acad. Sci. U.S.A.*, 80:6071–6075.
9. Johnson, B.E., Nau, M.M., Gazdar, A.F., Carney, D.N., Oie, H.K., Minna, J.D., and Ihde, D.C. (1985): Oncogene amplification of c-*myc* and N-*myc* in cell lines established from patients with small cell lung cancer (SCLC) is associated with shortened survival. *Proc. ASCO,* 4:186.
10. Kohl, N.E., Kanda, N., Schreck, R.R., Bruns, G., Latt, S.A., Gilbert, F., and Alt, F.W. (1983): Transposition and amplification of oncogene-related sequences in human neuroblastomas. *Cell,* 35:359–367.
11. Kohl, N.E., Legouy, E., DePinho, R.A., Nisen, P.D., Smith, R.K., Gee, C.E., and Alt, F.W. (1986): Human N-*myc* is closely related in organization and nucleotide sequence to c-*myc*. *Nature,* 319:73–77.
12. Land, H., Parada, L.F., and Weinberg, R.A. (1983): Tumorigenic conversion of primary embryo fibroblasts requires at least two cooperating oncogenes. *Nature,* 304:596–602.

13. Lee, W-H, Murphee, A.L., and Benedict, W.F. (1984): Expression and amplification of the N-*myc* gene in primary retinoblastoma. *Nature,* 309:458–460.
14. Little, C.D., Nau, M.M., Carney, D.N., Gazdar, A.F., and Minna, J.D. (1983): Amplification and expression of the c-*myc* oncogene in human lung cancer cell lines. *Nature,* 306:194–196.
15. Michitsch, R.W., Montgomery, K.T., and Melera, P.W. (1984): Expression of the amplified domain in human neuroblastoma cells. *Mol. Cell. Biol.,* 4:2370–2380.
16. Morton, C.C., Diamond, A., Lane, M.A., Cooper, G.M., and Leder, P. (1984): Mapping of the human B-lym-1 transforming gene activated in Burkitt lymphomas to chromosome 1. *Science,* 223:173–175.
17. Nau, M.M., Brooks, B.J., Battey, J., Sausville, E., Gazdar, A.F., Kirsch, I.R., McBride, O.W., Bertness, V., Hollis, G.F., and Minna, J.D. (1985): L-*myc*: A new *myc*-related gene amplified and expressed in human small cell lung cancer. *Nature,* 318:69–73.
18. Nau, M.M., Brooks, B.J., Carney, D.N., Gazdar, A.F., and Minna, J.D. (1986): Human small cell lung cancers with amplification and expression of the N-*myc* gene. *Proc. Natl. Acad. Sci. U.S.A., (in press).*
19. Ryan, J., Barker, P.E., Shimizu, K., Wigler, M., and Ruddle, F.H. (1983): Chromosomal assignment of a family of human oncogenes. *Proc. Natl. Acad. Sci. U.S.A.,* 80:4460–4463.
20. Ravetch, J.M., Siebenlist, U., Korsmeyer, S., Waldmann, T., and Leder, P. (1981): Structure of the human immunoglobulin in μ locus: Characterization of embryonic and rearranged J and D genes. *Cell,* 27:583:591.
21. Sausville, E., Carney, D., and Battey, J. (1985): The human vasopressin gene is linked to the oxytocin gene and is selectively expressed in a cultured lung cancer cell line. *J. Biol. Chem.,* 260:10236–10241.
22. Schmike, R.T. (1982): In: *Gene Amplification,* edited by R.T. Schmike, pp. 317–333. Cold Spring Harbor Laboratory, Cold Spring Harbor, New York.
23. Schmike, R.T., Brown, P.C., Kaufman, R.J., McGrogan, M., and State, D.L. (1981): Chromosomal and extrachromosomal localization of amplified dihydrofolate reductase genes in cultured mammalian cells. *Cold Spring Harbor Symp. Quant. Biol.,* 55:785–797.
24. Schwab, M., Alitalo, K., Klempnauer, K.-H., Varmus, H.E., Bishop, J.M., Gilbert, F., Brodeur, G., Goldstein, M., and Trent, J. (1983): Amplified DNA with limited homology to *myc* cellular oncogene is shared by human neuroblastom cell lines and a neuroblastoma tumour. *Nature,* 305:245–248.
25. Schwab, M., Varmus, H.E., and Bishop, J.M. (1985): Human N-*myc* gene contributes to neoplastic transformation of mammalian cells in culture. *Nature,* 316:160–162.
26. Schwab, M., Varmus, H.E., Bishop, J.M., Grzeschik, K.-H., Naylor, S.L., Sakaguchi, A.Y, Brodeur, G., Trent, J. (1984): Chromosome localization in normal human cells and neuroblastomas of a gene related to c-*myc*. *Nature,* 308:288–291.
27. Seeger, R.C., Brodeur, G.M., Sather, H., Dalton, A., Siegel, A.E., Wong, K.Y., and Hammond, D. (1985): Association of multiple copies of the N-*myc* oncogene with rapid progression of neuroblastomas. *New Eng. J. Med.,* 313:1111–1116.
28. Silverberg, E. (1985): Cancer statistics. *Ca-A Cancer Journal for Clinicians,* 35:19–35.
29. Simms, E., Gazdar, A.F., Abrams, P.G., and Minna, J.D. (1980): Growth of human small cell (oat cell) carcinoma of the lung in serum-free growth factor-supplemented medium. *Cancer Res.,* 40:4356:4363.
30. Taub, R., Kirsch, I., Morton, C., Lenoir, G., Swan, D., Tronick, S., Aaronson, S., and Leder, P. (1982): Translocation of the c-*myc* gene into the immunoglobulin heavy chain locus in human Burkitt lymphoma and murine plasmacytoma cells. *Proc. Natl. Acad. Sci. U.S.A.,* 79:7837–7841.
31. Whang-Peng, J., Bunn, P.A., Kao-shan, C.S., Lee, E.C., Carney, D.N., Gazdar, A.F., Minna, J.D. (1982): A nonrandom chromosomal abnormality, del 3p(14–23), in human small cell lung cancer (SCLC). *Cancer Genet. Cytogenet.,* 6:119–134.
32. Wurster-Hill, D.H. and Maurer, L.H. (1978): Cytogenetic diagnosis of cancer: Abnormalities of chromosomes and polyploid levels in the bone marrow of patients with small cell anaplastic carcinoma of lung. *J. Natl. Cancer Inst.,* 61:1065–1075.
33. Yancopoulos, G.D., Nisen, P.D., Tesfaye, A., Kohl, N.E., Goldfarb, M.P., and Alt, F.W. (1985): N-*myc* can cooperate with *ras* to transform normal cells in culture. *Proc. Natl. Acad. Sci. U.S.A.,* 82:5455–5459.

Burkitt's Lymphoma, A Human Cancer Model for the Study of the Multistep Development of Cancer: Proposal for a New Scenario

*Gilbert M. Lenoir and **Georg W. Bornkamm

*International Agency for Research on Cancer, 69372 Lyon Cedex 08, France; and
**Institut für Virologie im Zentrum für Hygiene,
7800 Freiburg-i.B., Federal Republic of Germany

Burkitt's lymphoma (BL) was recognized more than 25 years ago as a distinct pathoclinical entity with peculiar epidemiological characteristics (22,24,97). The tumor, which occurs at high incidence in certain areas of Central Africa, was suspected of having a viral etiology, and malaria was also proposed as an important risk factor (23,60). This hypothesis was strengthened by the isolation of the Epstein-Barr virus (EBV) from a BL culture (40), identification of viral DNA sequences within the tumor cells (159), and by the demonstration that EBV could immortalize human B lymphocytes *in vitro* and induce lymphoproliferative disorders when injected into subhuman primates (57). At that stage, BL might have been considered to be mainly a virally induced cancer. However, we have learned since that the development of BL is a multistep process in which EBV is not the sole transforming factor. Additional cellular changes must occur in malignant BL cells in order to explain the monoclonality of the tumor and the phenotypic differences between EBV genome-containing BL cells and EBV-immortalized non-malignant lymphoid cells (39,67).

It has become increasingly apparent that certain nonrandom chromosomal changes that are characteristic of malignant BL cells might represent a critical step in the development of the malignant clone. The finding that the chromosomal rearrangements involve the locus of a specific cellular proto-oncogene (c-*myc*) (140) and may lead to its deregulation has opened new avenues for studying interactions of viral and cellular genes in the genesis of a cancer (67,68). Burkitt's lymphoma is one of the rare human cancers for which a multistep scenario has been proposed based on epidemiological and biological evidence, and which has been related to molecular changes characteristic of the steps by which the tumor evolves (69,70).

In this chapter we shall review the evidence supporting a multistep scenario, focusing on the EBV-associated BL that occurs in areas of Africa. In view of new knowledge gained from molecular studies, from studies of AIDS/BL cases and of a BL animal model, mouse plasmacytoma (MPC) (107), we propose that the sequence of events leading to the appearance of a Burkitt's cell be seriously reconsidered.

BURKITT'S LYMPHOMA

The main features of this cancer, in which the significance of a viral association and chromosomal translocations has been studied so extensively during the past two decades, can be summarized as follows (see also ref. 76, for review).

Burkitt's Lymphoma Is a Well-Defined Pathoclinical Entity

The tumor now called BL was initially described as a distinct pathoclinical entity occurring at high frequency in children in Central Africa. Cytologically and pathologically, the initial description is still valid: BL is a malignant lymphoma comprising a monomorphic outgrowth of undifferentiated lymphoid cells, with little variation in size and shape, an amphophilic cytoplasm with clear vacuoles, and a noncleaved nucleus containing two to five basophilic nucleoli (97,149). At low magnification, a "starry-sky" pattern is frequently observed, caused by macrophage infiltration of this rapidly growing tumor (7). In the new working formulation of non-Hodgkin's lymphoma, BL belongs to the group of high-grade malignant lymphomas, with small noncleaved cells (127). BL consists of a clonal proliferation of lymphocytes from the B-cell subset, all of which synthesize heavy-chain immunoglobulins (Ig), predominantly of the μ subtype (109). Immunoglobin light-chain expression is also observed in most but not all BL tumors. The cells never fully differentiate to a plasmocytic phenotype.

Burkitt's lymphoma occurs predominantly in young people; the peak incidence is in children between 4 and 12 years old. The tumor is more common in males with a male:female ratio of 2:1 to 3:1. Jaw tumors are frequent in African children, whereas abdominal masses represent the most frequent tumors at presentation in children from other areas (105).

Burkitt's Lymphoma Incidence Shows Very Wide Geographic Variations

On the basis of the histological and cytological criteria for defining BL described above, this tumor occurs throughout the world (99) and is one of the most frequent (if not the most frequent) cancers in African children (82). Outside Africa (in Europe and the United States), it represents about 3% of childhood cancers (for review, see refs. 78 and 82); however, it represents a significant proportion of malignant non-Hodgkin's lymphomas, i.e., from 30% to 40% in all regions. In high-incidence areas of Africa, BL occurs in 5 to 15 children/100,000/ year; the estimated incidence in western countries (low-incidence areas) is 20- to 100-fold lower (35). There is still a lot of uncertainty about the incidence of BL in areas such as South America, Asia, and North Africa, because of large variations in diagnostic practices. In Africa, the distribution of the tumor is markedly dependent on geographicoclimatic factors, such as temperature, rainfall, and altitude. Several epidemiological characteristics of African BL can be explained on the basis of its relationship to malaria infection (60,91,98). the major characteristics of the malaria/BL association are *(a)* a strong geographical correlation between the incidence of BL and the intensity of *Plasmodium falciparum* transmission; *(b)* a close correlation between the age incidence of BL and the age of acquiring maximum levels of antimalarial Ig; and *(c)* relative protection from BL with residence in urban areas, where levels of malaria transmission are lower, compared with rural areas. Furthermore, the incidence of BL is elevated in areas with both a high rate of parasite infection and high parasitemia (hyperholoendemic regions). This might explain why the incidence of BL is relatively low in some populations of Latin America and Southeast Asia, which have high malaria rates but low parasitemia. The geographic distribution of BL also led to the suggestion that in Africa, the disease might be due to virus spread by an insect vector (24).

Association of Epstein-Barr Virus with Burkitt's Lymphoma Is Important but not Consistent

The search for a viral etiology led to the isolation of EBV from a BL cell line. Subsequently, DNA of this ubiquitous human herpes virus (which infects latently 95–98% of the adult population throughout the world) was detected in malignant cells of the great majority (96%) of BL cases from high-incidence areas (45). However, in the remaining 4% and in up to 85% of cases occurring in certain low-incidence areas (Europe, United States), viral markers are not present in malignant BL cells (78). When these figures are compared to the incidence of BL inside and outside high-risk areas, they suggest that the incidence of EBV-negative BL in equatorial Africa is very similar to the overall incidence of BL in Europe or North America, i.e., approximately 0.2 to 0.5/100,000/year. This is consistent with the possibility that EBV genome-negative BL occurs at similar incidence throughout the world and that the main variation in BL incidence is linked mainly to variations in the incidence of EBV genome-positive BL. However, our finding that in North Africa (72), a region that cannot be considered to be a high-incidence area, the EBV/BL association is as strong as in Central Africa, indicates that the relationship between EBV/BL association and BL incidence is more complex than anticipated and deserves further attention.

Burkitt's Lymphoma Cells Can Be Grown in Tissue Culture

One of the characteristic features of BL, which was of great assistance in laboratory investigations on this cancer, is the relative ease with which malignant cells can be cultivated *in vitro* (41,81). Using simple tissue culture procedures, most

BL tumor cells can be established as permanent cell lines—an almost unique phenomenon in the field of human cancer. Most of the major advances in laboratory studies of BL etiology were made using BL cell lines: the initial discovery of EBV (40), the first description of the $14q^+$ chromosomal marker (85), and, more recently (140), molecular studies implicating c-*myc* oncogene transposition.

It was thought for some time that the possibility of establishment of BL cell lines was limited to EBV-associated cases. Our experience with tumors from low-incidence areas such as France (81), where most BL do not contain the EB viral genome, shows that malignant cells can almost always be grown in culture. This property can thus be considered to be a characteristic feature of Burkitt-type lymphomas, independent of their association with EBV. This feature, remarkable for a human cancer, appears to be relatively independent of exogenous factors, fetal calf serum being the only major requirement.

Importance of Chromosomal Translocation

Burkitt's lymphoma cells are characterized by specific translocations. Cytogenetic studies on BL that were carried out before the introduction of banding techniques gave conflicting results, and no specific chromosomal markers (for review, see ref. 9), were identified. In 1972, however, at a time when most research was devoted to the etiological role of EBV in BL etiology, G. Manolov and Y. Manolov (85) used the newly developed banding techniques to demonstrate that BL cells are characterized by an abnormal chromosome 14, which has an extra band at the extremity of its long arm: $14q^+$. This work was rapidly confirmed by others and was expanded by Zech et al. (156) and Kaiser-McCaw et al. (61) who demonstrated that the extra material on chromosome 14 originated from the long arm of chromosome 8. The t(8;14) translocation was thus discovered. Another important fact was noted by these two groups: The cytogenetic anomaly was also observed in non-EBV-associated BL, suggesting that the translocation was a characteristic marker of the tumor, independent of its association with the virus.

In 1979, a study of cases originating from so-called nonendemic areas (now better defined as low-incidence areas) led to a new, important cytogenetic discovery of BL cases without the $14q^+$ marker but with two new translocations, designated *variant* translocations. The t(2;8) was initially described in cases from Japan, Belgium, and the Federal Republic of Germany (20,90,145). In France, meanwhile, Berger et al. (11) reported one case of BL with a t(8;22) translocation, followed immediately by three further cases. The important feature was that in all three translocations—t(8;14) ($q24;q32$), t(2;8) ($p11;q24$), and t(8;22) ($q24;q11$)—the breakpoint on chromosome 8 was cytogenetically the same: $8q24$. Soon after, BL cases with variant translocations were also reported from high-incidence areas, i.e., in Africa (13), indicating that independent of the geographic origin of the patient or the EBV-association of the tumor, the characteristic cytogenetic anomaly of BL is a translocation of chromosome 8 (band $q24$) with chromosome 2, 14, or 22 (13,79). Cytogenetically, the translocations are considered to be reciprocal and balanced (86).

In order to estimate the relative frequency of the three types of BL translocation, we summarized the data obtained from 51 of our BL lines (15 African, 13 North African, and 23 Caucasian) (80); t(8;14) is by far the most frequent translocation observed in BL (76% of the cases), only one fourth of the tumors presenting with a variant translocation; t(8;22) and t(2;8) are not equally distributed, the former being at least twice as frequent as the latter.

How consistently are chromosomal translocations involving chromosome $8q24$ found in BL? On the basis of cytogenetic studies performed on BL cell lines, there is apparently no exception; more than 100 lines have been analyzed, and all carried one of the three characteristic translocations. One unique, extensively studied cell line derived from an EBV genome-negative African B-cell lymphoma, designated BJAB, was found to have no anomaly involving chromosome number 8 (156) and was considered to be an example of BL without the translocation. However, the diagnosis of BL in this case remains very uncertain, on the basis of several criteria, such as clinical history, morphology, and surface phenotype. Normal karyotypes have also been reported after analysis of fresh BL; however, as pointed out by Berger et al. (10), this might be due to tumor infiltration by nonmalignant cells.

Are t(8;14), t(8;22), and t(2;8) specific to BL? In his review, Mitelman (89) indicated that t(8;14) has also been found in non-Burkitt malignant lymphoproliferations. This is, however, not frequent. The study of Sigaux et al. (132) indicates that on the basis of morphometric measurements of malignant cells, rearrangement of $8q24$ with chromosome 14 is found mainly in lymphomas that show a morphological continuum extending from BL to immunoblastic lymphomas. They might in fact all correspond to cell proliferation of the B-cell lineage.

Epstein-Barr virus does not appear to be capable of inducing BL translocations directly. No such chromosomal rearrangement is observed in lymphoid cells during infectious mononucleosis *in vivo*. Furthermore, lymphoblastoid cell lines (LCL) obtained *in vitro* by immortalization of nonmalignant lymphocytes have normal karyotypes at the time of their establishment; when they are kept in culture even for several years, the chromosomal changes that are observed are never BL-type translocations (131). In BL patients, the translocations have only been observed in malignant cells and never in other hematopoietic cell types such as stem cells. The association is thus very different from that of t(9;22) or Ph^1 with chronic myelogenous leukemia (CML), in which it has been shown that Ph^1 may be *(a)* an early event in the disease process, observed, for example, before clinical evidence of CML in atomic bomb survivors; and *(b)* detected in other cell lineages, such as nonmalignant B lymphoid cells (14), indicating that the t(9;22) is most probably already present in stem cells.

MOLECULAR CONSEQUENCES OF THE TRANSLOCATIONS

We cannot review here all of the molecular characteristics of BL translocations so far studied. An excellent review has just been prepared, in which the original contributions are cited (28). We shall take the mechanistic point of view and describe here the important features of the c-*myc* gene and of the translocations, to lay the basis for a discussion of various possible scenarios for BL pathogenesis. Two points are particularly important: (a) What is the role of the c-*myc* gene in normal proliferation? *(b)* Can structural analysis of the breakpoints tell us anything about the time when the translocation occurs during B-cell development?

Involvement of the c-*myc* Locus Is a Characteristic Feature of Burkitt's Lymphoma

For the first time, the genes on both loci participating in a human chromosomal translocation have been identified. In fact, the genes and their gene products were already known: the Ig genes on chromosomes 14, 2, and 22, and the c-*myc* gene on the long arm of chromosome 8 (75).

The c-*myc* gene was originally defined as the cellular homologue of the gene transduced by avian oncogenic retroviruses (for review, see ref. 16). It is a member of a family of related genes, including N-*myc* and L-*myc*, which have frequently been implicated in oncogenesis in various species in association with chromosomal translocation, promoter or enhancer insertion, and gene amplification. The gene consists of one noncoding and two coding exons. Transcription is initiated at two sites in exon 1. The gene codes for two closely related proteins of molecular weights between 62 and 68 kilodaltons (kD) located in the nucleus, the biochemical function of which is not known (28).

The c-*myc* locus has been found to be implicated in all BL cases, but in different ways in the different types of translocation. As indicated in Fig. 1, in t(8;14) translocations c-*myc* moves from chromosome 8 to chromosome 14. The c-*myc* breakpoints are always located upstream of the c-*myc* coding region, either in the first intron, the first exon, or upstream thereof, strictly avoiding damage to the c-*myc* coding region. The translocation fuses the c-*myc* gene, usually at the μ-switch site, to an Ig heavy-chain constant gene, in a head-to-head fashion. The Ig heavy-chain enhancer is usually separated from the heavy-chain constant gene and is present on the reciprocal fragment of the translocation, unavailable for the c-*myc* gene (28,30,75,80).

In the variant t(2;8) and t(8;22) translocations, c-*myc* does not usually present with a detectable rearrangement and remains on chromosome 8. However, the Ig light-chain loci move to chromosome 8 (Fig. 1) such that c-*myc* is again located 5' of an Ig constant region, now in a head-to-tail orientation (28,30,80). Remarkably, the distance between the c-*myc* gene and the translocation (and hence the Ig light chain involved) is usually considerable and may vary from a few kilobases (kb) (33,58) to more than 50 kb (M. Lipp and G. Hartl, *personal communication*) or perhaps more, in view of the analogy with variant translocation in MPC (29). In all cases with variant translocations that have been studied in detail, mutations have been observed in the noncoding first exon of the c-*myc* gene (33,111,113,141).

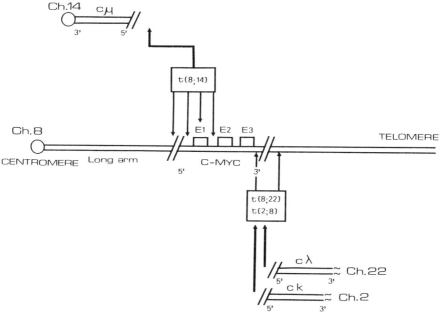

FIG. 1. Diagrammatic representation of the translocation breakpoint with respect to the c-*myc* gene in the three types of translocations. The three c-*myc* exons are represented by boxes.

The translocation and c-*myc* gene involvement are neither a tissue culture artifact nor a constitutional anomaly, since they are found in the original tumor as well as in the cell line established from the tumor and are absent from the patient's nonmalignant lymphoid cells (Fig. 2).

Role of the *myc* Gene in Cellular Proliferation

In order to define the function of c-*myc* and the translocation in a multistep scenario, we summarize briefly what we know about the roles of c-*myc* in normal growth regulation and of *myc*-containing retroviruses in oncogenesis. After mitogenic stimulation of resting lymphocytes or on addition of serum to serum-deprived fibroblasts, c-*myc* is expressed maximally after 1 to 2 hr (65,66,94). The expression of c-*myc* is preceded by the expression of c-*fos*, which is stimulated maximally after 15 to 20 min (48, 94). These observations suggest that c-*fos* and c-*myc* operate in the pathway that converts a signal from outside the cell into a proliferative response.

More information about the role of the c-*myc* gene in this pathway was provided by experiments that made it possible to define further the early events that take

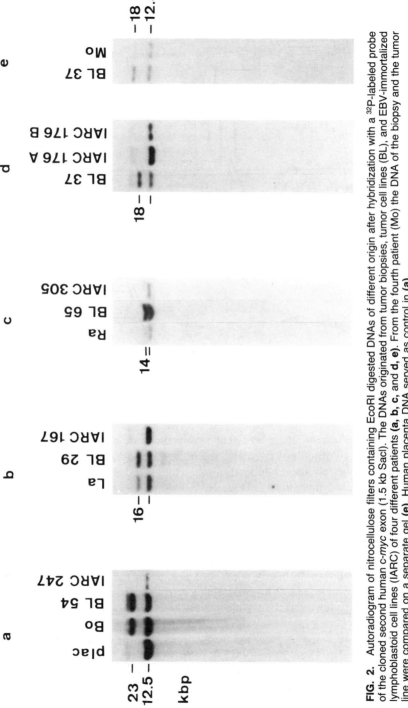

FIG. 2. Autoradiogram of nitrocellulose filters containing EcoRI digested DNAs of different origin after hybridization with a ^{32}P-labeled probe of the cloned second human c-myc exon (1.5 kb SacI). The DNAs originated from tumor biopsies, tumor cell lines (BL), and EBV-immortalized lymphoblastoid cell lines (IARC) of four different patients (a, b, c, and d, e). From the fourth patient (Mo) the DNA of the biopsy and the tumor line were compared on a separate gel (e). Human placenta DNA served as control in (a).

place when cells are stimulated to proliferate. Thus, c-*myc* expression releases cells from their requirement for one class of growth factors (defined as competence factors) and renders them responsive to another (progression factors) (3,59,117,138). The c-*myc* protein can thus be regarded as a competence factor, the expression of which is required but is not sufficient for cell proliferation.

A similar competence-progression model holds true for the growth control of T and B lymphocytes. The first step in lymphocyte activation consists in interaction of the antigen receptor with an antigen (for review, see ref. 88). Binding of antigen is prerequisite for lymphocytes to respond to growth factors. During B- and T-cell activation, antigen can be replaced by mitogens such as Sepharose-bound anti-μ antibodies and by lectins such as PHA or ConA. Mitogenic stimulation thus leads to a manifold increase in c-*myc* expression and makes the cells responsive to the action of growth factors (65,66,119).

Expression of c-*myc* is required not only at one particular time point in the cell cycle, but also for the sequential and regulated appearance of receptors (96,120), since their expression is required for progression through the cycle and for entrance into the next G1 phase (rather than G0). This may account for the invariant expression of c-*myc* throughout the cell cycle (54,104,114,142).

A second possible approach to defining the role of *myc* has been to study its effect following transduction by retroviruses, DNA-mediated transfection, or introduction of the *myc* gene into the germ line of mice. The central question in each approach is whether the *myc* gene is sufficient for tumor induction or whether additional events are required. The second possibility is favored by a number of observations. Retroviruses containing not only *myc* but also additional oncogenes, such as *raf/mil*, induce tumors after a significantly shorter latent period than viruses containing only the *myc* gene alone (6,118). This suggests that the effect of the *myc* oncogene can be complemented by those of other genes for the rapid development of a fully malignant phenotype. Furthermore, avian cells transformed by a virus containing the *myc* gene only are not tumorigenic unless they spread the virus (15,103,129). Secondary infection by the virus of a large number of cells may facilitate the selection for further cellular events required for tumorigenicity.

In DNA-mediated transfection experiments on rat embryo fibroblasts, *myc* induces continuous proliferation of the culture and reduces the requirement for growth factors (64,73,92). The cells still show contact inhibition and are nontumorigenic. In this cellular system, *myc* gives the cells the capacity to form foci in response to epidermal growth factor (123) or tumor promoters (27) and to be transformed easily by other oncogenes, such as *ras* and Py large T (73). Thus, *myc* cooperates with other factors to render the primary cells fully malignant.

Studies of transgenic mice (1,137) indicate that a constitutionally active c-*myc* gene can be a 100% risk factor for cancer in these animals. Thus, *myc* is a true cancer gene. However, the animals survive long enough to develop a monoclonal tumor surrounded by apparently normal cells (1), indicating that additional events have occurred in the tumor cell clone to make the *myc*-expressing cells fully malignant.

How Do the Immunoglobin Loci Affect the Regulation of the Translocated c-*myc* Gene?

This question is difficult to answer, since the physiological modes of c-*myc* regulation are poorly understood at the molecular level. Therefore, we summarize only briefly some important features of c-*myc* regulation.

Expression of the c-*myc* gene is dependent on a number of factors that regulate the steady-state amount of c-*myc* protein at different levels inside the cell. The c-*myc* gene is transcribed at a high rate (31,48), c-*myc* RNA and c-*myc* protein are extremely short-lived (15–30 min) (31,53,116), and translational control may also be important (32). The c-*myc* gene thus appears to be closely controlled at any given level in the cell.

The steady-state level of c-*myc* RNA in individual BL lines varies by a factor of at least 5 from cell line to cell line. It is about the same or slightly higher than in EBV-immortalized cells (52,83,141). In mouse plasmacytomas, the amount of steady-state c-*myc* RNA does not exceed that in normal lymphocytes stimulated with mitogens (65), suggesting that constitutive expression rather than overexpression is the important determinant. In BL and MPC cells c-*myc* is always derived from the chromosome affected by the translocation, whereas the normal allele appears to be transcriptionally silent, suggesting that transcriptional regulation is important (28).

Another indication of transcriptional deregulation is the alteration in utilization of the two c-*myc* promoters that is observed in BL and MPC. In nonmalignant fibroblasts and EBV-immortalized cells, the second of these promoters is used preferentially, whereas in BL and MPC, there is a shift in promoter usage in favor of the first (141,150). Since the first exon is noncoding, and transcripts initiated at the first and second promoter have the same turnover and are thought to code for the same protein, the consequence of the shift in promoter usage is not yet understood.

The following modes of deregulation have been proposed: transcriptional deregulation by truncation of the c-*myc* gene or loss of transcriptional feedback control as a consequence of mutations in or around the first exon (37b,75,111), transcriptional deregulation due to translocation to or insertion of an immunoglobulin gene enhancer (28,56), increased stability of c-*myc* RNA due to structural alteration of the truncated message (38,106,112), increased translatability of the c-*myc* message due to the alteration in the secondary structure (32), and alteration of the c-*myc* protein by mutation (113). A change in the c-*myc* coding region may have some importance in some cases; however, there is good evidence that such changes are not required (5,113,136,148).

Alternatively to several different mechanisms, one of which is decisive in each individual tumor, all BL may involve a common mechanism of deregulation. Thus, heretofore unrecognized enhancer elements in or downstream of the Ig constant regions might operate over a long distance, as proposed by Croce and Nowell

(30). Destruction of an important regulatory region in the c-*myc* gene by truncation or mutation might be a second general mechanism (28).

Another intriguing possibility pointed out by Adams et al. (1) is that activation of the c-*myc* gene may be an indirect effect of the Ig enhancer, even if c-*myc* and the enhancer are located on different chromosomes. The Ig enhancer might be required for initiation of an active chromatin configuration but not for its maintenance. This suggestion is based on two kinds of observation: first, that a c-*myc* gene fused to the Ig enhancer and introduced into the mouse germ line induces a high rate of B cell lymphomas in the offspring (1); and second, that Ig genes remain actively transcribed *in vivo* even when the enhancer has been lost by deletion (146,155).

How Does the Translocation Occur?

The first question is obviously whether EBV induces chromosomal aberrations, including the specific translocations of BL. Evidence against this possibility has already been described above.

The important question for the multistep scenario for BL is whether the translocation occurs in a differentiating B cell that is in the process of rearranging its Ig genes or, alternatively in a mature B cell that has already accomplished the Ig rearrangement. Can the analysis of the translocation breakpoint provide an answer to this question? Two different types of Ig gene rearrangements have to be discriminated. The first type, VDJ and VJ joining in heavy- and light-chain loci, gives rise to the formation of functional Ig genes necessary to the generation of mature B cells with a capacity to respond to antigen. The second type, class switch, occurs in a B cell and gives rise to the change of Ig isotype without changing the antigenic specificity and involves specific sequence elements located in front of heavy-chain constant gene as a target (2,144). Class switch is believed to be induced by antigen in a mature B cell carrying functional VDJ and VJ rearrangements .

The breakpoints of the translocations on the Ig genes are usually located at or close to sites that are also used for one or the other types of physiological Ig rearrangement (28). In most BL cases with the t(8;14) translocation, they are located within the μ switch region or close to the joining region. In most MPC, they are located within the α switch region.

In variant translocations involving the light-chain loci, the breakpoints are also localized at or close to the joining regions, which are the targets for the physiological light-chain gene rearrangement. If we interpret the translocation as an accident of physiological Ig rearrangement, this would imply that at least in those cases in which joining regions are involved, the translocation would occur during early B-cell differentiation before the cell has acquired the capacity to respond to antigen.

Does involvement of μ switch sites in most cases with t(8;14) indicate that the translocation occurred in a mature B cell that is stimulated to switch by the action of antigen? This is an obvious possibility. It is, however, also conceivable that the translocation occurred before a functional heavy chain gene is formed, the μ switch region just being a favored site of recombination. Two essential requirements for recombination at the μ switch site are also fulfilled in pre-B cells: *(a)* the site is accessible to the switch recombinase or other enzymes by transcriptional activation [apparently also required for VDH joining (2)]; *(b)* the switch recombinase is active also in pre-B cells, as shown by the possibility of class switching in Abelson virus-transformed pre-B cell lines (for review, see ref. 2).

Is the involvement of the α switch site in most MPC also compatible with the translocation occurring at an early stage of B-cell differentiation, such as during VDJ rearrangement? Yes, for two possible reasons. In lymphocytes with the propensity to produce IgA, the switch to cα might already occur during VDJ rearrangement. Alternatively, since the switch induced by antigenic stimulation usually involves the functional as well as the nonfunctional allele (115), c-*myc* might move from switch μ to switch α as a secondary event.

All BL and MPC cells express one functionally rearranged heavy-chain gene, suggesting that heavy-chain expression is a prerequisite for the cells to survive and to develop into a tumor. This phenomenon is most easily explained by assuming that a cell without functional rearrangement, regardless of whether c-*myc* has invaded the Ig locus or not, will continue to rearrange its Ig locus until a functional Ig has been formed. Otherwise, it will be eliminated like any other B-cell without functional rearrangement (see also ref. 71).

It is particularly important to consider the generation of the translocation in the kinetic context of B-cell differentiation. The turnover of B cells is extremely high: they are newly generated at such a high rate that the whole B-cell repertoire is renewed within days (102). This steady-state equilibrium, generating new cells and cell death, is disturbed only when a given B cell encounters its appropriate antigen. Then, an expanding clone gives rise to primary, and eventually to secondary, immune response.

From studies of the functional role of c-*myc* in signal transduction, we have learned that c-*myc* expression can abrogate the requirement for signals that provide competence for growth, such as platelet-derived growth factor in fibroblasts or antigenic stimulation in lymphocytes. Activation of c-*myc* by the translocation may thus mimic the effect of antigens and may prevent the cell from decaying, like B cells that did not encounter their respective antigen do.

In the BL and MPC studied, only c-*myc* or the corresponding chromosomal region but no other oncogenes have been found juxtaposed to the Ig loci. The c-*myc* gene may therefore carry target structures that are recognized preferentially and used by the Ig rearranging machinery; however, no evidence for the occurrence of such sequences has been obtained (12). It is conceivable that an open chromatin configuration of the c-*myc* locus related to the enormous proliferating capacity of the progenitor cells (102) favors the involvement of the c-*myc* locus. Selection for these Ig-c-*myc* translocations may also be an important contributing factor. Under the specific environmental conditions related to BL and MPC, perhaps only cells with translocations affecting c-*myc* and one of the Ig loci can expand and later become malignant.

Are Cellular Oncogene Systems Other than c-*myc* Involved in Burkitt's Lymphoma Pathogenesis?

This question has been raised in the context of elucidating the genesis of non-EBV-associated BL. As discussed above, in many animal systems, the activated *myc* gene is usually not capable of inducing a fully transformed phenotype, and other genes are frequently required (73). In EBV-associated BL, EBV may cooperate with the activated c-*myc*, although this has not yet been demonstrated. What occurs in EBV-negative BL? In two cases, activated *ras* genes were identified by transfection into NIH 3T3 cells: A mutated N-*ras* gene was detected in the long-established Ramos line (95), and a mutated ki-*ras* was isolated in a fresh BL tumor obtained from a patient in relapse (77). This finding indicates that a translocated *myc* gene and a mutated *ras* gene can occur simultaneously in a B cell, but this cannot yet be considered to occur in all BL cases. The role of the B-*lym* gene, suggested to cooperate with c-*myc* in EBV-positive and -negative BL, is still controversial (37). Using a cytogenetic approach, Berger and Bernheim (8) have suggested that anomalies of the long arm of chromosome 1 occur at higher frequency in EBV-negative than in EBV-positive BL. This suggests that those anomalies and the presence of the EBV genome may share some functions pertaining to cell proliferation. Taken together, these observations indicate that no other major candidate oncogene has been identified in BL with high frequency. However, the activation of genes acting in a complementary manner with c-*myc* might well be required, especially in EBV-negative BL. In at least some cases, these genes might belong to the *ras* family.

EPSTEIN-BARR VIRUS

The association between EBV and BL is neither direct nor simple. In order to facilitate the discussion of the importance of EBV in BL pathogenesis, we have summarized some of the main biological and molecular characteristics of this virus and of the host-virus relationship (see also refs. 39,57).

Epstein-Barr Virus Is a Ubiquitous Virus Associated with Two Human Malignancies

Epstein-Barr Virus is associated with two human cancers—not only BL, from which it was first isolated, but also nasopharyngeal carcinoma (NPC), an undifferentiated carcinoma highly prevalent in southern Chinese populations. In the two cases, the association is based both on serological evidence and on formal identification of viral markers (DNA or antigens) within malignant cells. Despite its well-documented association with these two forms of cancer, EBV is also ubiquitous: It infects populations in all parts of the world and is not restricted to populations at high risk for BL or NPC. Infection of human populations by this virus is largely asymptomatic (57).

Primary Epstein-Barr Virus Infection Can Cause Infectious Mononucleosis

In most communities, primary infection with EBV occurs during the first few years of life and is clinically silent. However, when infection is delayed until the second decade or later, as occurs in many countries with high socioeconomic levels, it may be accompanied in a small proportion of cases with the clinical symptoms of infectious mononucleosis (IM), a self-limited lymphoproliferative disease. Infectious mononucleosis is characterized in the very early phase by proliferation of nonproductively infected B lymphocytes, followed rapidly by a strong reaction of lymphoid T cells to limit the proliferation. T cells then dominate the white blood cell population, giving rise to the designation "mononucleosis." Transplacental infection with EBV has never been documented, and it is assumed that the main way in which EBV is transmitted is through close salivary contact. All primary infections, in whatever form, induce permanent seroconversion to viral antibody positivity and the establishment of a lifelong carrier state whereby the virus persists in some not yet fully identified forms.

Epstein-Barr Virus has a Dual Tropism for B Lymphocytes and Epithelial Cells

B lymphocytes represent the target in the blood for EBV, which attaches itself to a complement receptor (CR2). *In vitro* viral infection is followed by immortalization of B lymphocytes into permanent LCL. The proliferating cells are infected latently, and only a very small proportion sustain active viral replication, indicating that the B cells are capable of controlling, at least partly, expression of the replicative functions of EBV. No *in vitro* cellular system that is fully permissive for viral replication is presently available. There is evidence *in vivo,* however, that the virus that is excreted in the saliva is produced in epithelial cells of the oropharynx (133). The demonstration in AIDS patients of viral replication in the epithelial cells of the lateral part of the tongue suggests that this might be a natural site of viral replication (49). This situation is reminiscent of Marek's disease virus, a chicken herpes virus that replicates in epithelial cells of the feather follicles, infects T lymphocytes nonproductively, and causes fatal proliferation of T cells *in vivo*.

How is B Cell Proliferation Controlled *in vivo*?

Multiple host-cell responses act to control EBV infection, and the production of specific anti-EBV antibodies is probably of critical importance for neutralization of infectious virions (122). Control of latently infected B cells is, however, mainly under the control of lymphoid T-cell-mediated activities. During primary EBV infections, such as IM, most of the cells that act against the proliferating EBV-infected B cells do so by antibody-dependent cytotoxicity and natural-killer-like mechanisms. However, an EBV-specific HLA immune-restricted cytotoxic response is rapidly induced (88a,121) that permits long-term control of EBV infections in seropositive individuals. The antigen recognized by these cytotoxic T cells on the surface of infected B lymphocytes has been defined as EBV-associated lymphocyte-detected membrane antigen (LYDMA) but has yet to be biochemically identified.

How Is the Carrier State Defined?

The number of virus-carrying lymphoid cells in peripheral blood has been estimated to be as high as one in 2,000 during IM, and one in 10^5 to 10^7 in normal individuals (124,152). The ease with which spontaneously outgrowing cell lines can be established from most EBV seropositive individuals probably reflects the steady state between continuous infection of B lymphocytes in the lymphoid tissue of the oropharynx and killing of infected B cells by the immune system. This view of the virus carrier state is based on the observation that excretion of virus in the saliva and the ability to establish spontaneous cell lines from peripheral blood are closely linked (151,152). The steady-state equilibrium could be perturbed by one of two mechanisms—an increase in virus production, which would increase the number of infected cells, or more important, an impairment of lymphoid T-cell control.

Epstein-Barr Virus Can Cause Lymphoproliferative Diseases in Individuals with Immune Dysfunctions

The importance of immune surveillance is stressed by the observation that individuals with immunodeficiencies may develop uncontrolled proliferation of EBV-immortalized lymphoid cells. Most such lymphoproliferations are polyclonal B cell proliferations, classified as diffuse lymphoma but not of the Burkitt's type. Such fatal proliferation can occur during primary infection (IM); it is very rare in the general population but is a frequent cause of death in children with genetically determined immunodeficiencies, such as X-linked lymphoproliferative syndrome (XLP) (110).

Epstein-Barr virus-associated lymphoproliferative disorders can also occur at relatively high incidence in seropositive individuals (already latently infected with EBV) who are treated with immunosuppressive therapy for organ transplantation (55). The implication of EBV in those lymphomas is based on the detection of viral markers in the proliferating cells. The importance of alterations to immune functions in the genesis of these tumors is stressed by the fact that they may regress when the immunosuppressive therapy is reduced or withdrawn (136a).

Is the Molecular Basis of Epstein-Barr Virus-Induced Cell Proliferation Understood?

Many research groups have used molecular and genetic approaches to identify the EBV genes that trigger B cell proliferation and are involved in the establishment of the latent state, in the hope of elucidating the role of EBV in malignant disease.

Epstein-Barr Virus is a very complex DNA virus of the group of herpesviruses (4). Comparison of the genomes of viruses originating from patients with different EBV-associated diseases provided no evidence for the existence of disease-specific substrains (18,21). In both BL and NPC cells, as well as in lymphocytes immortalized by EBV, the viral genome is present in multiple (10–50) copies as circularly closed episomal DNA (63,81a). Integration of a single viral DNA copy into the host genome has also been reported to occur, but not consistently (87). The significance of integration is still unknown.

A Limited Number of Genes is Regularly Transcribed in Epstein-Barr Virus-Immortalized Cells

One of the major transcripts, a 3.7-kb RNA, which corresponds to the Bam HI-K fragment of the EBV genome, codes for the major EBV nuclear antigen (EBNA 1) (139). It has been shown that EBNA binds specifically to the origin of plasmid replication of EBV and is required for maintenance of the viral DNA episomal state (118a,153,154).

A second nuclear antigen, EBNA 2, has been identified that is coded by the Bam H1-Y-H region (37a,56a,93,130). Even though the biochemical function of EBNA 2 is unknown, there is good evidence that it is required for the initiation of immortalization by EBV. The P3HR-1 EBV isolate that has a deletion for the EBNA 2 gene (19,58c) has lost its immortalizing capacity (44,133a).

A 2.5-kb RNA transcribed from the right end of the EBV genome codes for latent membrane protein (42,58b,84). Its structure, deduced from its DNA sequence, suggests that it is a membrane protein that might serve as an ion-transport molecule. Because of its membrane localization, this protein has been suggested to be LYDMA, the virally encoded membrane antigen that is recognized by cytotoxic T-cells. There is, however, no direct evidence to support this hypothesis.

Whether EBNA 2 and latent membrane protein are involved in inducing the cell proliferation remains to be shown.

The pattern of viral RAN in immortalized cells has been shown to be more complex than previously regarded (17,57a,134). It is thus very likely that additional proteins cooperate with EBNA 1, EBNA 2, and the latent membrane protein in inducing the latent viral state and cell immortalization.

Cell immortalization probably occurs by EBV-mediated activation of a physiological B cell activation pathway (143). Experiments by Gordon et al. (46,47) suggest that at least some of the immortalizing action of EBV is due to an autocrine loop: EBV-immortalized B cells release B-cell growth factor, and their continuous proliferation depends on it (see also ref. 135). It is possible that this phenomenon is also mediated by activation of specific cellular oncogenes, as suggested also by the observation that the *fgr* oncogene seems to be transcribed specifically in EBV-infected cells (26).

THE MULTISTEP SCENARIO

In formulating a multistep scenario for BL development, two points must be taken into consideration:

1. It is necessary to make a clear distinction between the events that occurred at the cellular level, at the the host level, and at a population level (epidemiology). This is particularly relevant for the sequence of events: In the natural history of African BL, primary EBV infection of the host may be an early event (34), followed by the appearance of cellular clone(s) having specific chromosomal translocation; at the cellular level, however, as discussed later, the translocation may preceed a subsequent EBV infection of (a) genetically altered cell(s) which in turn give(s) rise to the tumor clone.

2. One must avoid generalization and refer only to specific situations. For instance, the two-stage model of transformation of rat embryo fibroblasts by the oncogenes such as *myc* and *ras* has revealed the importance of cooperation between oncogenes (73). In this context, *myc* has been defined as an immortalizing, nontransforming gene, and this classification has been useful under specific experimental conditions (74). Attempts to generalize the concept of cooperating immortalizing and transforming genes, however, have generated confusion and created the apparent paradox that two immortalizing functions (*myc* and EBV) would have to cooperate in BL. Furthermore, in the context of retroviruses, it is not acceptable to classify *myc* as a nontransforming gene (6,16). Its important function is rather to render cells highly susceptible to (transforming) growth factors, and there is thus no reason why *myc* should not cooperate with EBV in BL.

Epstein-Barr virus-associated BL does not occur as a simple and direct consequence of impairment of the immune control of a latent viral infection. In congenitally or iatrogenically immunosuppressed patients, this type of situation leads to the development of EBV genome-positive lymphomas, classified as large-cell or immunoblastic lymphomas rather than BL (57,110). They are usually polyclonal in origin and composed of proliferating cells lacking the specific BL translocations. Thus, BL cannot be considered the direct consequence of reactivation of latent EBV infection by the immune deficient state; the situation is more complex. However, the information gained from the study of malignant lymphomas occurring in AIDS patients may help to elucidate the pathways leading either to polyclonal EBV-associated lymphoproliferative diseases or to BL.

The Lesson of AIDS

There is growing evidence that male homosexuals are at increased risk of acquiring malignant lymphomas, mainly aggressive non-Hodgkin's lymphomas (62,157). These lymphomas are predominantly high-grade B-cell neoplasms classified as diffuse large-cell lymphomas, but also as small, noncleaved, lymphoblastic and true Burkitt-type lymphomas (carrying the specific translocations) (25,62,146a,158). The diffuse large-cell lymphomas are very similar to those occurring in other immunodeficient individuals and occur in AIDS patients with severely altered immune parameters (62). They represent polyclonal proliferation and have usually been EBV-associated. The Burkitt-type tumors usually occur in those patients who have previously had generalized reactive lymphadenopathy and in whom the immune dysfunction is less severe (62). This suggests that in the latter patients, the T-cell functions are still capable of controlling EBV-induced polyclonal B-cell proliferation. However, multiple factors that trigger B-cell subsets, such as antigenic stimuli associated with concurrent infections, could lead to lymphoid hyperplasia or generalized lymphadenopathies and eventually to the development of a true monoclonal neoplasia. Thus, the risk factor for developing BL appears not to be the strong immunosuppression that permits the proliferation of "normal" EBV infected B cells, but the LAV/HTLVIII immune alteration, which leads to multiple antigenic stimulation. Emergence of the malignancy may be favored by lymphoid stimulation and, at a relatively late stage, by EBV infection. This scenario is very different from the EBV-induced proliferation that occurs in fatal infectious mononucleosis or in renal transplant recipients. It indicates the importance of the triggering of the B-cell subset by complex mechanisms involving microbial or parasitic infection. The critical role of immune stimulation seen in BL associated with AIDS may also occur in the genesis of African BL, with malaria infection as the major B-cell stimulus.

The Lesson of African Burkitt's Lymphoma

Epidemiologically, two causal factors—malaria and EBV—have been implicated in the pathogenesis of African BL. Hyperholoendemic malaria is the only factor so far identified that is common to those regions of the world in which BL occurs at a high incidence (60,91,98). Furthermore, in these regions, there is a close correlation between the age of occurrence of the clinical BL and age of acquiring maximum Ig levels and antimalarial antibodies. The evidence linking malaria to the induction of BL indicates that the relationship is not simply with parasitic infection, but rather with the host immune response to intense and prolonged parasitic load (98). In such regions, malaria imposes a continuing and intense lymphoid stimulation. The highly increased production of Ig (5- to 6-fold) that is triggered by parasitic infection must reflect in a considerable increase in B-lymphocyte turnover, in addition to increased stimulation of the entire reticuloendothelial system (50,51,128).

How is EBV causally associated with BL? All children in the high-incidence areas are infected very early in life, 90% usually before the age of 2 years (34). The large prospective study conducted by IARC in Uganda, showed that high antibody titers to EBV capsid antigens before onset of disease represent a major risk indicator for BL in children (36). However, we do not know whether this serological pattern reflects a massive infection at an early age in the children at risk or a strong serological response due to other reasons. Is there an absence of cellular immune control of the latent EBV infection in this high-risk group? *In vitro* studies have demonstrated that acute malaria might affect EBV-specific T-cell cytotoxicity (91a,147); however, clinical studies indicate that in Africa, primary EBV infection is usually silent, IM not often recognized, and there has been no report of fatal IM nor of EBV-related polyclonal lymphoproliferative disorders in these populations, suggesting that the cellular immune control has not been overridden (57).

In comparing the pathogenesis of EBV-associated BL in African children and in AIDS patients, we realize that an important common feature might be not lack of control of the latently EBV-infected B-cell population, but rather an intense triggering of the B-cell immune system (98), by either malaria or opportunistic infections. In this respect, pristane induction of granuloma is critical in the genesis of MPC, an animal model highly analogous to BL (see M. Potter, *this volume*); and pristane has been shown to produce a burst in the production of B lymphocytes by a central effect on the bone marrow (43).

The Lesson of Molecular Studies

Generation of the Translocation Appears to Require an Active Immunoglobin Recombinational Machinery

Molecular identification of the site of involvement of the Ig loci in the BL chromosomal translocation suggests that the mechanisms involved in Ig recombinations (VDJ as well as switching) are also probably involved in the genesis of the translocation with c-*myc*. Therefore, the translocation should appear during immunocyte maturation, when the recombinational machinery is operating. This can occur at an early stage when committed stem cells start making VDJ rearrangements at the μ locus, during light-chain rearrangement or during heavy-chain switching. All experimental data are compatible with the view that the translocation occurs early during B-cell differentiation when functional Ig genes are formed (2,28,100).

myc Expression is Required but not Sufficient for B-cell Growth

The experimental studies described above provide good evidence that *myc* expression is not sufficient for cell proliferation. *myc* activation seems to represent a critical step, making cells competent and responsive to growth-promoting signals. For the establishment of a fully malignant phenotype, the cooperation of *myc* with other (onco)genes might be required.

What Might Be the New Scenario for Burkitt's Lymphoma Development?

Can we try to formulate another scenario for the genesis of BL that will take into account the epidemiological, experimental, and molecular information that has become available during the past few years? The BL that occurs in high-incidence areas of Africa can be used as a model.

Klein has suggested in 1979 (67) that BL develops in at least three steps, and proposed the following sequence of events (Table 2):

1. The first step is *EBV-induced immortalization* of B lymphocytes on primary infection.

2. The second step is *stimulation of the proliferation of EBV-carrying B cells*. The probability of genetic aberrations increases with the number of cell divisions. This step is facilitated in the presence of malaria through B-cell triggering and through suppression of T cells involved in the control of the proliferation of EBV-infected B cells.

3. The third and final step is the *reciprocal translocation that leads* to *myc* deregulation, to the development of a malignant clone, and to the appearance of a tumor mass (see also refs. 69–71).

This model has been highly stimulating and useful for generating new hypotheses. It stressed the importance of a second cellular event in the genesis of BL, which was subsequently identified as the c-*myc* gene translocation. However, in view of what has been learned during the past several years, this sequence of events is unlikely. Molecular studies, the plasmacytoma model, and the lesson of AIDS all suggest strongly that, at the cellular level, the translocation might be the first event (J. H. Béchet, *personal communication;* 101) and that infection of this cell with EBV occurs at a second stage, even though the primary EBV infection in the patient may have occurred long before the emergence of any cytogenetically abnormal cell *in vivo*.

What are the data that suggest that the sequence of events proposed in 1979 is improbable?

TABLE 1. *Comparison of two scenarios for BL development*

	Step 1	Step 2	Step 3
Present scenario[a]	B-cell immortalization	Increase in the size of the target-cell population	*myc* Deregulation Malignant transformation
In Africa due to:	EBV: associated with early primary infection	Malaria: parasite-induced T-cell immunosuppression as well as polyclonal B-cell activation	Chromosomal translocation *myc*/Ig juxtaposition: probability increased by increased number of cell divisions
Proposed scenario	Triggering of B-cell system	Appearance of "competent" cells by *myc* deregulation	Autonomously growing malignant cell
In Africa due to:	Malaria-induced polyclonal B-cell activation: increases Ig recombinational events per time	Chromosomal translocation *myc*/Ig juxtaposition: makes B-cells responsive to growth signals in the absence of antigenic stimulation	EBV: infection of a cell carrying the translocation (infection and outgrowth favored by immune dysfunction)

[a]From ref. 67.

1. The occurrence of a BL-type translocation in an EBV-immortalized proliferating cell is very unlikely: Latently, EBV-infected proliferating B-cells have already functionally rearranged Ig genes, and Ig recombinational activity must therefore be very low in these cells. Immonuglobulin loci are not preferential sites of chromosomal breaks in LCL (131), even if the cells are kept proliferating in tissue culture for many years.

2. The second step in the model of G. Klein (expansion of EBV infected cells *in vivo*) is based on two experimental or clinical assumptions, for which we have no evidence: first, that there are sufficient EBV-infected cells in the body with the capacity to proliferate under the conditions of the virus-host balance that has developed in the presence of holoendemic malaria; second, that these cells proliferate *in vivo* at such a high rate that the risk of acquiring the translocation is significantly increased.

In EBV-seropositive individuals, there are indeed many EBV-infected cells in the blood (as many as 10^7?) (152); however, their proliferating capacity has never been documented, except under conditions in which there is no balance between virus and host, i.e., during primary infection and during EBV-induced lymphoproliferation (122). As pointed out earlier, the presence of EBV-positive cells in the blood probably represents the steady-state equilibrium between new infection of lymphocytes and elimination of infected cells by immune-competent T-cells and does not imply that proliferation-competent EBV-infected cells are present in the blood (122).

The second assumption is that immune dysfunction due to malaria is the precipitating factor that increases the chances for the malignant clone to evolve; however, even though T-cell immune functions may be diminished in African children (148), they are still effective enough to establish a virus-host balance. Thus, EBV-associated polyclonal lymphoproliferative disorders have not been reported to occur in these populations. From analysis of the studies of lymphomas in AIDS patients, we came to the same conclusion—that severe T-cell dysfunction is not *the* important risk factor for BL.

We propose an alternative model for BL development. The following scenario also assumes that BL develops in at least three steps, but in a different sequence (Table 2):

1. The first step is the *generation of lymphoid cells at high risk for occurrence of translocations involving the Ig loci*. This is achieved by chronic, persistent immunological stimulation of the B-cell subset by holoendemic malaria or AIDS-associated opportunistic infections, superimposed on the extremely high physiological turnover of B cells (about 5×10^7/day in mice) (102). Such immunological stimulation has an experimental counterpart in the pristane-induced granuloma required for the development of MPC (107,108).

2. The second step is the appearance of the *translocation involving c-myc and Ig loci* in one or a few B cells. This is the first genetic event. The translocation alone is not sufficient to transform cells to a neoplastic state, but the first molecular event has made the lymphoid cells competent and responsive to external growth factors (cell proliferation is still dependent on these growth factors). These cells are not immortal and will usually die unless the third event occurs.

3. The third step is *infection* by EBV of a B cell carrying the specific translocation. This represents the second genetic event. The acquisition of viral functions renders the cell independent of exogenous growth factors and thus autonomous for its growth. The third step could also be achieved by a second cellular event, such as activation of another oncogene, and this may be the case in most EBV-negative BL occurring in low-incidence areas. This event may occur, however, at a very low frequency. Is a fourth step (or more) required? This cannot be ruled out, even though *myc* deregulation through translocation plus EBV should represent the two critical cellular events for rendering a cell malignant.

In EBV-associated BL, however, the production of a tumor mass can be achieved only if the T-cell immune system that normally controls the proliferation of infected B cells does not kill the EBV-infected malignant cells. This could be brought about by impairment of that cellular activity or, as suggested by the experiments of Rooney et al. (125,126), by the rapid *in vivo* selection of a subclone with altered lymphocyte-detected membrane antigen (LYDMA) expression from the expanding population of proliferating malignant cells. It is also conceivable that in some cases, the EBV-specific T-cell immune system is merely overridden by cells with an enormous proliferative capacity.

As opposed to the previous model, which involved, briefly; (1) virus, (b) proliferation of EBV-infected B-cells by T-cell suppression and B-cell stimulation, and (c) chromosomal translocation, we now propose (a) triggering of the immune B-cell system, (b) chromosomal translocation, and (c) complementation by the virus. Although this sequence of events is different, the two critical steps remain the same: chromosomal translocation and virus infection of the cell. However, our model reinforces the role of holoendemic malaria and assumes that triggering of parasite-induced B cells is a critical step. In G. Klein's model, the increased probability of *myc*/Ig juxtaposition is due to an increased number of divisions of lymphoid cells, based on the assumption that "the probability of all genetic aberrations increases in direct relationship with the number of cell divisions." It seems to us that the risk of aberrant genetic recombination involving the Ig loci in a B cell is related mainly to the Ig recombinational activity in these cells rather than to their division. As a consequence, the risk of *myc*/Ig juxtaposition will be increased by any factor that increases the number of Ig rearrangements over time (by increasing either the number of cells rearranging Ig genes or by increasing their turnover and/or their recombinational activity). The two models also differ in the importance of EBV functions: According to G. Klein's model, they might no longer be necessary in a cell carrying the translocation, whereas in our model, EBV functions are critical for complementing *myc* activity.

Holoendemic malaria is characterized by marked hypergammaglobulinemia, reflecting extensive activation of host B lymphocytes (50,128). A very limited proportion of the Ig produced is directed against malaria antigenic determinants, suggesting that the parasite possesses nonspecific or polyclonal activating properties (51,128). The massive generation of B cells that are rearranging their Ig gene should represent a major risk factor for the appearance of a cell carrying a BL specific translocation. Our model favors the triggering of *nonvirally infected* cells, since expansion of EBV-infected B cells, as proposed by G. Klein, is controlled negatively by T-cell immunity.

In view of present knowledge about B-cell proliferation in mice and humans and about the function of the *myc* gene, it is tempting to speculate that cells containing the *myc*/Ig translocation become competent and can respond to growth factors. Additionally, *myc* activation may be equivalent to antigenic stimulation, not only in making the cell responsive to growth factors, but also in preventing its immediate decay. A prolonged half-life may be an important factor in increasing the probability of the subsequent step. However, such cells are not malignant and do not even have the capacity to proliferate unless exogeneous growth factors are present.

Epstein-Barr virus infection should liberate cells from their environment by furnishing functions usually provided by growth factors. It has been proposed that

EBV immortalizes B lymphocytes by inducing autocrine stimulation (46,47). This viral function might well complement the *myc* gene function in making the cell malignant in the same way that *ras* and *myc* genes complement each other in transforming rat embryo fibroblasts. As suggested by zur Hausen (160), *myc* and EBV might not only cooperate in making the cells autonomous for growth, but also in rendering them unresponsive to growth-inhibiting signals, which *in vivo* repress cell growth. Such a mechanism could account for the fact that BL and EBV-immortalized cells differ in their capacity to grow in nude mice *in vivo*, whereas their growth behavior *in vitro* is very comparable.

The step of viral infection might be critical to explain the high incidence of BL in Africa, since it is possible that most of the cells carrying the translocation will die unless they become infected. In Central Africa, the probability of virus infection of a cell carrying the translocation (in other words, the availability of infectious virus within the organism), might represent a critical risk factor.

Viruses are shed continuously in the oropharynx from most, if not all, EBV-seropositive individuals in close contact with the lymphoid tissue of Waldeyer's ring, resulting in continuous infection of new B cells (122,152). Increased virus shedding thus increases the number of infected B cells and may represent an important risk factor. High virus production in the oropharynx may be related to a very early age of primary infection and/or to immune impairment associated with malaria. High virus production, related to a large number of infected lymphocytes and, as a consequence, a high rate of elimination of infected cells by the immune system, is probably reflected in high-antibody titers against virion components. The results of the IARC prospective study in Africa, showing that children at risk had high antiviral capsid antigen titers long before development of BL (36), are in perfect agreement with this hypothesis. Direct evidence that patients at risk for BL, i.e., AIDS patients, have dramatically increased virus shedding has also been provided by Greenspan et al. (49).

In this new scenario, EBV still represents a major component: Viral genes are required to complement the action of c-*myc* in the cell. At the epidemiological level, viral infection could still represent a major contributing factor together with malaria accounting for the elevated incidence of BL in Africa. This model also agrees with that proposed by Ohno et al. (101) for MPC induced by pristane and Abelson virus, in which viral infection is also very likely to represent the second cellular event and greatly accelerates the occurrence of tumors.

Does our model also apply to EBV-negative BL occurring in low-incidence areas and to MPC in the absence of Abelson virus? In MPC the pristane-induced granuloma appears to play a dual role. On the one hand, it may induce B-cell stimulation in the marrow (43). On the other hand, by providing growth factors in the inflamed tissue, it may select for cell seeding into the granuloma and be responsive to these growth factors. In EBV-negative BL, a granuloma equivalent has not been identified. Epidemiological studies might identify risk factors associated with the stimulation of the B-cell system, such as microbial or other parasite infections. If no risk factor is identified, it may be that the occurrence of EBV-negative BL reflects the natural occurrence of abnormal Ig recombination during immunological development.

Could proliferation of a cell carrying the translocation be initiated by interaction with its specific antigen? This seems to be conceivable. Antigen-driven proliferation might select for secondary changes such as activation of other oncogenes. Constitutive c-*myc* expression may also prevent terminal differentiation and render the cell insensitive to feedback control mechanisms that limit the immune response physiologically. With regard to step 3, we may assume that in EBV-negative BL, activation of other cellular genes capable of cooperating with *myc*, must occur. Their identification and mode of activation remain to be elucidated.

CONCLUSIONS AND PERSPECTIVES

BL has long been used as a cancer model in fields as different as epidemiology, virology, immunology, cytogenetics, molecular biology, and clinical studies (see, for review ref. 76). This cancer is also one of the first for which the hypothesis that tumors arise through accumulation of several changes affecting cell growth could be dissected in biological and molecular terms.

On the basis of the data available at the end of the 1970s, G. Klein was able to formulate a stimulating hypothesis, which stressed the importance of two cellular events in the genesis of BL: infection by EBV and the chromosomal translocation. Testing of this hypothesis was instrumental in the identification of the molecular consequences of the chromosomal anomaly: dysregulation of a cellular oncogene through its transposition to another locus.

The sequence of events proposed at that time: first virus, then chromosomal translocation, was strongly influenced by the fact that from the time EBV was identified in 1964, BL was considered to be a virally induced cancer. Presently, however, on the basis of new knowledge about the biology of the virus, B-lymphocyte differentiation, and molecular aspects of the translocation, we propose that the sequence of cellular events leading to the genesis of BL should be reversed—the virus infects a cell already carrying the *myc*/Ig translocation.

Several implications result from this different order of events. Since, in our model, the triggering of B cells by malaria or by opportunistic infections in AIDS patients represents a major risk factor, it will be important to learn more about the kinetic aspects of B-cell turnover under these conditions. The MPC model of M. Potter *(this volume)*, which has been extremely useful in the past, may prove to be ideal for studying this question experimentally. In view of the pathophysiological similarity of BL and MPC, it will be most important to study the histologically identifiable pre-lesions in which the tumor cells evolve. Again, the MPC model may provide the experimental tool for identifying at which stage the translocation occurs and what are the growth factors present in the granuloma that support expansion of a premalignant clone. With regard to humans, AIDS patients may well represent a unique human population in which BL pre-lesions could be identified and studied.

As far as the role of EBV is concerned, our model stresses the importance of identifying viral functions capable of cooperating with *myc* in making the B cell malignant, which are probably those required for viral latency and immortalization. The risk of BL should be related to the likelihood that a cell carrying the translocation will be infected. This potentially important risk factor might be estimated quantitatively by determining the numbers of cells in the blood of high- and low-risk individuals who have been infected by EBV and can release the virus *in vitro*. High virus shedding might also represent an important risk factor for EBV-associated polyclonal lymphoproliferative disorders. If this proves to be correct, therapeutic interference with virus production in the oropharynx might be possible.

According to the model proposed here, a second cellular event plays an important role in the development of EBV-negative BL. The involvement of other oncogenes or another virus is conceivable. Antigen-driven proliferation might allow selection for secondary changes. Epidemiological studies may help to identify risk factors for BL in areas other than Africa and for BL in AIDS patients; however, in view of the low frequency of the disease, such studies might be difficult to carry out and the outcome questionable. More knowledge about normal B-cell growth control and about EBV-induced proliferation may guide us to other cellular genes, the involvement of which cannot be foreseen on the basis of present knowledge.

We hope that the model presented here will stimulate discussions and research in this and related fields and will help to generate new and testable hypotheses.

ACKNOWLEDGMENTS

We would like to thank R. Berger, D. Eick, J. P. Lamelin, M. Lipp, G. O'Connor, M. Pawlita, A. Polack for helpful suggestions, and C. Fuchez and E. Heseltine for their expertise in preparing and editing the manuscript.

REFERENCES

1. Adams, J.M., Harris, A.W., Pinkert, C.A., Corcoran, L.M., Alexander, W.S., Cory, S., Palmiter, R.D., and Brinster, R.L. (1985): The c-*myc* driven by immunoglobulin enhancers induces lymphoid malignancy in transgenic mice. *Nature*, 318:533–538.
2. Alt, F.W., Blackwell, T.K., DePinho, R.A., Reth, M.A., and Yancopoulos, D. (1986): Regulation of genome rearrangement events during lymphocyte differentiation. In: *Immunological Reviews: Control of Immunoglobulin Gene Expression*, edited by G. Møller, Vol. 89, pp. 5–30. Stockholm.
3. Armelin, H.A., Armelin, M.C.S., Kelly, K., Stewart, T., Leder, P., Cochran, B.H., and Stiles C. (1984): Functional role for c-*myc* in mitogenic response to platelet derived growth factor. *Nature*, 310:655–660.
4. Baer, R., Bankier, A.T., Biggin, M.D., Deininger, P.L., Farrell, P.J., Gibson, T.J., Hatfull, G., Hudson, G.S., Satchwell, S.C., Séguin, C., Tuffnell, P.S., and Barrell, B.G. (1984): DNA sequence and expression of the B95-8 Epstein-Barr virus genome. *Nature*, 310:207–211.
5. Battey, J., Moulding, C., Taub, R., Murphy, W., Stewart, T., Potter, H., Lenoir, G.M., and Leder, P. (1983): The human c-*myc* oncogene: structural consequences of translocation into the IgH locus in Burkitt lymphoma. *Cell*, 34:779–787.

6. Bechade, C., Calothy, G., Pessac, B., Martin, J. Coll, Denhez, F., Saule, S., Ghysdael, J., and Stehelin, D. (1985): Induction of proliferation or transformation of neuroretina cells by the *mil* and *myc* viral oncogenes. *Nature,* 316:559–562.
7. Berard, C.W., O'Conor, G.T., Thomas, L.B., and Torloni, H. (ed.) (1969): Histopathological definition of Burkitt's tumour. *Bull. W.H.O.,* 40:601–607.
8. Berger, R. and Bernheim, A. (1984): Is there any functional equivalence of chromosome 1 long arm abnormalities to Epstein-Barr virus presence in Burkitt's Lymphoma cell lines? *C.R. Acad. Sci. Paris,* 6:143–145.
9. Berger, R., and Bernheim, A. (1985): Cytogenetics of Burkitt's lymphoma-leukaemia: A review. In: *Burkitt's Lymphoma: A Human Cancer Model,* edited by G.M. Lenoir, G. O'Conor, and C.L.M. Olweny, IARC Scientific Publications No. 60, pp. 65–80. IARC, Lyons.
10. Berger, R., Bernheim, A., Daniel, M.T., Valensi, F., and Flandrin, G. (1983): Cytological types of mitoses and chromosome abnormalities in acute leukemia. *Leuk. Res.,* 7:221–236.
11. Berger, R., Bernheim, A., Weh, H.J., Flandrin, G., Daniel, M.T., Brouet, J.C., and Colbert, N. (1979): A new translocation in Burkitt's tumor cells. *Hum. Genet.,* 53:111–112.
12. Bernard, O., Cory, S., Gerondakis, S., Webb, E., and Adams, J.M. (1983): Sequence of the murine and human cellular *myc* oncogenes and two modes of *myc* transcription resulting from chromosome translocation in B-lymphoid tumors. *EMBO J.,* 2:2375–2383.
13. Bernheim, A., Berger, R., and Lenoir, G. (1981): Cytogenetic studies on African Burkitt's lymphoma cell lines t(8;14), t(2;8) and t(8;22). *Cancer Genet. Cytogenet.,* 3:302–315.
14. Bernheim, A., Berger, R., Preud'homme, J.L., Labaume, S., Bussel, A., and Barot-Ciorbaru, R. (1981): Philadelphia chromosome positive blood B-lymphocytes in chronic myelocytic leukemia. *Leuk. Res.,* 5:331–339.
15. Beug, H., von Kirchbach, A., Doderlein, G., Conscience, J.F., and Graf, T. (1979): Chicken haematopoetic cells transformed by seven strains of defective avian leukaemia viruses display three distinct phenotypes of differentiation. *Cell,* 18:375–390.
16. Bister, K., and Jansen, H.W. Oncogenes in Retroviruses and cells: biochemistry and molecular genetics. *Adv. Cancer Res. (in press).*
17. Bodescot, M., Chambraud, B., Farrell, P., and Perricaudet, M. (1984): Spliced RNA from the IR1-U2 region of Epstein-Barr virus: presence of an open reading frame for a repetitive polypeptide. *EMBO J.,* 3:1913–17.
18. Bornkamm, G.W., Delius, H., Zimber, U., Hudewentz, J., and Epstein, M.A. (1980): Comparison of Epstein-Barr virus strains of different origin by analysis of the viral DNAs. *J. Virol.,* 35:603–618.
19. Bornkamm, G.W., Hudewentz, J., Freese, U.K., and Zimber, U. (1982): The deletion of the non-transforming Epstein-Barr virus strain P3HR-1 causes fusion of the large internal repeat to the DS_L region. *J. Virol.,* 43:952–968.
20. Bornkamm, G.W., Kaduk, B., Kachel, G., Schneider, U., Fresen, K.O., Schwanitz, G., and Hermanek, P. (1980): Epstein-Barr virus positive Burkitt's lymphoma in a German woman during pregnancy. *Blut,* 40:167–177.
21. Bornkamm, G.W., von Knebel-Doeberitz, M., and Lenoir, G.M. (1984): No evidence for differences in the Epstein-Barr virus genome carried in Burkitt Lymphoma cells and nonmalignant lymphoblastoid cells from the same patients. *Proc. Natl. Acad. Sci. USA,* 81:4930–14934.
22. Burkitt, D.P. (1958): A sacroma involving the jaws in African children. *Br. J. Surg.,* 46:218–223.
23. Burkitt, D.P. (1962): A children's cancer dependent on climatic factors. *Nature,* 194:232–234.
24. Burkitt, D.P. (1983): The discovery of Burkitt's lymphoma. *Cancer,* 51:1777–1786.
25. Chaganti, R.S.K., Jhanwar, S.C., Koziner, B., Arlin, Z., Mertelsmann, R., and Clarkson, B.D. (1983): Specific translocations characterize Burkitt's-like lymphoma of homosexual men with acquired immunodeficiency syndrome. *Blood,* 61:1269–1272.
26. Cheah, M., Ley, T., Tronick, S., and Robbins, K. (1986): fgr proto-oncogene mRNA induced in B-lymphocytes by Epstein-Barr virus infection. *Nature,* 319:238–240.
27. Connan, G., Rassoulzadegan, M., and Cuzin, F. (1985): Focus formation in rat fibroblasts exposed to a tumor promoter after transfer of polyoma *plt* and *myc* oncogenes. *Nature,* 314:277–279.
28. Cory, S. (1986): Activation of cellular oncogenes in Hemopoietic cells by chromosome translocation. *Adv. Cancer. Res. (in press).*
29. Cory, S., Graham, M., Webb, E., Corcoran, L., and Adams, J.M. (1985): Variant (6;15) translocations in murine plasmacytomas involve a chromosome 15 locus at least 72 kb from the c-*myc* oncogene. *EMBO J.,* 4:675–681.
30. Croce, C.M., and Nowell, P.C. (1985): Molecular basis of human B-cell neoplasia. *Blood,* 65:1–7.

31. Dani, C., Blanchard, J.M., Piechaczyk, M., El Sabouty, S., Marty, L., and Jeanteur, P. (1984): Extreme instability of c-*myc* mRNA in normal and transformed cells. *Proc. Natl. Acad. Sci. U.S.A.,* 81:7046–7050.
32. Darveau, A., Pelletier, J., and Sonenberg, N. (1985): Differential efficiencies of *in vitro* translocation of mouse c-*myc* transcripts differing in the 5' untranslated region. *Proc. Natl. Acad. Sci. U.S.A.,* 82:2315–2319.
33. Denny, C.T., Hollis, G.F., Magrath, I.T., and Kirsch, J.R. (1985): Burkitt lymphoma cell line carrying a variant translocation creates new DNA at the breakpoint and violates the hierarchy of immunoglobulin gene rearrangement. *Mol. Cell. Biol.,* 5:237–254.
34. de Thé, G. (1977): Is Burkitt's Lymphoma related to a perinatal infection by Epstein-Barr virus? *Lancet,* 1:335–338.
35. de Thé, G. (1985): Epstein-Barr virus and Burkitt's Lymphoma worldwide: the causal relationship revisited. In: *Burkitt's Lymphoma: A Human Cancer Model,* edited by G.M. Lenoir, G. O'Conor, and C.L.M. Olweny, IARC Scientific Publications No. 60, pp. 165–176. IARC, Lyons.
36. de Thé, G., Geser, A., Day, N.E., Tukei, P.M., Williams, E.H., Beri, D.P., Smith, P.G., Dean, A.G., Bornkamm, G.W., Feorino, P., and Henle, W. (1978): Epidemiological evidence for a causal relationship between Epstein-Barr virus and Burkitt's lymphoma: Results of a Ugandan prospective study. *Nature,* 274:756–761.
37. Diamond, A., Cooper, G.M., Ritz, J., and Lane, A. (1983): Identification and molecular cloning of the human Blym transforming gene activated in Burkitt's lymphomas. *Nature,* 305:112–116.
37a.Dillner, J., Kallin, B., Klein, G., Jörnvall, H., Alexander, H., and Lerner, R. (1985): Antibodies against synthetic peptides react with the second Epstein-Barr virus-associated nuclear antigen. *EMBO.,* 4:1813–1818.
37b.Dunnick, W., Shell, B.E., and Dery, C. (1983): DNA sequences near the site of reciprocal recombination between a *c-myc* oncogene and an immunoglobulin switch region. *Proc. Natl. Acad. Sci.,* 80:7269–7273.
38. Eick, D., Piechaczyk, M., Henglein, B., Blanchard, J.M., Traub, B., Kofler, E., Wiest, S., Lenoir, G.M., and Bornkamm, G.W. (1985): Aberrant c-*myc* RNAs of Burkitt's lymphoma cells have longer half-lives. *EMBO J.,* 4:3717–3725.
39. Epstein, M.A., and Achong, B.G. editors, (1979). In: *The Epstein-Barr Virus.* Springer Verlag, New York.
40. Epstein, M.A., Achong, B.G., and Barr, Y.M. (1964): Virus particles in cultured lymphoblasts from Burkitt's lymphoma. *Lancet,* 1:702–703.
41. Epstein, M.A., and Barr, Y.M. (1964): Cultivation in vitro of human lymphoblasts from Burkitt's malignant lymphoma. *Lancet,* 1:252–253.
42. Fennewald, S., van Santen, V., Kieff, E. (1984): Nucleotide sequence of an mRNA transcribed in latent growth-transforming virus infection indicates that it may encode a membrane protein. *J. Virol.,* 51:411–419.
43. Fulop, G.M., and Osmond, D.G. (1983): Regulation of bone marrow lymphocyte production. *Cell. Immunol.,* 75:80–90.
44. Fresen, K.O., Cho, M.S., and zur Hausen, H. (1980): Recombination between Epstein-Barr genomes. Viruses in naturally occurring cancers. *Cold Spring Harbor Conf. Cell Proliferation,* 7:35–44.
45. Geser, A., Lenoir, G., Anvret, M., Bornkamm, G.W., Klein, G., Williams, E.H., Wright, D.H., and de Thé, G. (1983): Epstein-Barr virus markers in a series of Burkitt's lymphoma from the West Nile District of Uganda. *Eur. J. Cancer Clin. Oncol.,* 19:1394–1404.
46. Gordon, J., Amann, P., Rosén, A., Ernberg, J., Ehlin-Henriksson, B., and Klein, G. (1985): Capacity of B-Lymphocytic lines of diverse tumor origin to produce and respond to B-cell growth factors: A progression model for B-cell lymphomagenesis. *Int. J. Cancer,* 35:251–256.
47. Gordon, J., Ley, S.C., Melanmed, M.D., English, L.S., and Hughes-Jones, N.C. (1984): Immortalized B-lymphocytes produce B-cell growth factor. *Nature,* 312:145–147.
48. Greenberg, M.E., and Ziff, E.B. (1984): Stimulation of 3T3 cells induces transcription of the c-*fos* proto-oncogene. *Nature,* 311:438–442.
49. Greenspan, J.S., Greenspan, D., Lennette, E.T., Abrams, D.I., Conant, M.A., Petersen, V., and Freese, U.K. (1985): Epstein-Barr virus replicates within the epithelial cells of oral "hairy" leukoplakia, an AIDS associated lesion. *N. Engl. J. Med.,* 313(25):1564–1571.
50. Greenwood, B.M. (1974): Possible role of a B-cell mitogen in hypergammaglobulinaemia in malaria and trypanosomiasis. *Lancet,* 1:435–436.

51. Greenwood, B.M., and Vick, R.M. (1975): Evidence for a malaria mitogen in human malaria. *Nature,* 257:592–594.
52. Hamlyn, P.H., and Rabbitts, T.H. (1983): Translocation joins c-*myc* and immunoglobulin γ1 genes in a Burkitt lymphoma revealing a third exon in the c-*myc* oncogene. *Nature,* 304:135–139.
53. Hann, S.R., and Eisenman, R.N. (1984): Proteins encoded by the human c-*myc* oncogene: Differential expression in neoplastic cells. *Mol. Cell. Biol.,* 4:2486–2497.
54. Hann, S.R., Thompson, C.B., and Eisenman, R.N. (1985): c-*myc* oncogene protein synthesis is independent of the cell cycle in human and avian cells. *Nature,* 314:366–369.
55. Hanto, D.W., Frizzera, G., Purtilo, D.T., Sakamoto, K., Sullivan, J.L., Saemundsen, A.K., Klein, G., Simmons, R.L., and Najarian, J.S. (1981): Clinical spectrum of lymphoproliferative disorders in renal transplant recipients and evidence for the role of Epstein-Barr virus. *Cancer Res.,* 41:4253–4261.
56. Hayday, A.C., Gillies, S.D., Saito, H., Wood, C., Wiman, K., Hayward, W.S., and Tonegawa, S. (1984): Activation of a translocated human c-*myc* gene by an enhancer in the immunoglobulin heavy-chain locus. *Nature,* 307:334–340.
56a. Hennessy, K., and Kieff, E. (1985): A second nuclear protein is encoded by Epstein-Barr virus in latent infection. *Science,* 227:1238–1240.
57. Henle, W., and Henle, G. (1985): Epstein-Barr virus and human malignancies. In: *Advances in viral Oncology,* Vol. 5, edited by G. Klein, pp. 201–238. Raven Press, New York
58. Hollis, G.F., Mitchell, K.F., Battey, J., Potter, H., Taub, R., Lenoir, G.M., and Leder, P. (1984): A variant translocation places the lambda/Immuno-globulin genes 3' to the c-*myc* oncogene in Burkitt's lymphoma. *Nature,* 307:752–755.
58a. Hudson, G.S., Bankier, A.T., Satchwell, S.C., and Barrell, B.G. (1985): The short unique region of the B95-8 Epstein-Barr Virus genome. *Virology,* 147:81–98.
58b. Hudson, G.S., Farrell, P.I., and Barrell, B.G. (1985): Two related but differentially expressed potential membrane proteins encoded by the Eco-RI D het region of Epstein-Barr virus B95-8. *J. Virol.,* 53:528–535.
58c. Jeang, K.-T., and Haywards, S.D. (1983): Organization of the Epstein-Barr virus DNA molecule. III. Location of the P3HR-1 deletion junction and characterization of the NotI repeat units that form part of the template for an abundant 12-0-tetradecanoylphorbol-13-acetate-induced mRNA transcript. *J. Virol.,* 48:135–148.
59. Kaczmarek, L., Hyland, J.K., Watt, R., Rosenberg, M., and Baserga, R. (1985): Microinjected c-*myc* as a Competence Factor. *Science,* 228:1313–1315.
60. Kafuko, G.W., and Burkitt, D.P. (1970): Burkitt's lymphoma and malaria. *Int. J. Cancer,* 6:1–9.
61. Kaiser-McCaw, B., Epstein, A.L., Kaplan, H.S., and Hecht, F. (1977): Chromosome 14 translocation in African and North American Burkitt's lymphoma. *Int. J. Cancer,* 19:482–486.
62. Kalter, S.P., Riggs, S.A., Cabanillas, F., Butler, J.J., Hagemeister, F.B., Mansell, P.W., Newell, G.R., Velasquez, W.S., Salvador, P., Barlogie, B., Rios, A., and Hersh, E.M. (1985): Aggressive Non-Hodgkin's lymphomas in immunocompromised homosexual males. *Blood,* 66:655–659.
63. Kaschka-Dierich, C., Adams, A., Lindahl, T., Bornkamm, G.W., Bjursell, G., Klein, G., Giovanella, B.C., and Singh, S. (1976): Intracellular forms of Epstein-Barr virus DNA in human tumor cells in vivo. *Nature,* 260:302–306.
64. Keath, E.J., Caimi, P.G., and Cole, M.D. (1984): Fibroblast lines expressing activated c-*myc* oncogenes are tumorigenic in nude mice and syngeneic animals. *Cell,* 39:339–348.
65. Keath, E.J., Kelekar, A., and Cole, M.D. (1984): Transcriptional activation of the translocated c-*myc* oncogene in mouse plasmacytomas: similar RNA levels in tumor and proliferating normal cells. *Cell,* 37:521–528.
66. Kelly, K., Cochran, B.H., Stiles, C.D., and Leder, P. (1983): Cell-specific regulation of the c-*myc* gene by lymphocyte mitogens and platelet-derived growth factor. *Cell,* 85:603–610.
67. Klein, G. (1979): Lymphoma development in mice and humans: Diversity of initiation is followed by convergent cytogenetic evolution. *Proc. Natl. Acad. Sci. U.S.A.,* 76:2442–2446.
68. Klein, G. (1981): The role of gene dosage and genetic transposition in carcinogenesis. *Nature,* 294:313–318.
69. Klein, G. (1983): Specific chromosomal translocations and the genesis of B-cell-derived tumors in mice and men. *Cell,* 32:311–315.
70. Klein, G., and Klein, E. (1985): Evolution of tumours and the impact of molecular oncology. *Nature,* 315:190–195.

71. Klein, G., and Klein, E. (1985): *Myc*/Ig juxtaposition by chromosomal translocations: some new insights, puzzles and paradoxes. *Immunol. Today,* 6(7):208–215.
72. Ladjadj, Y., Philip, T., Lenoir, G.M., Tazerout, F.Z., Bendisari, K., Boukheloua, R., Biron, P., Brunat-Mentigny, M., and Aboulola, M. (1984): Abdominal Burkitt-type lymphomas in Algeria. *Br. J. Cancer,* 49:503–512.
73. Land, H., Parada, L.F., and Weinberg, R.A. (1983): Tumorigenic conversion of primary embryo fibroblasts requires at least two cooperating oncogenes. *nature,* 304:596–602.
74. Land, H., Parada, L.F., and Weinberg, R.A. (1983): Cellular oncogenes and multistep carcinogenesis. *Science,* 222:771–778.
75. Leder, P., Battey, J., Lenoir, G., Moulding, C., Murphy, W., Potter, H., Stewart, T., and Taub, R. (1983): Translocations among antibody genes in human cancer. *Science,* 222:765–771.
76. Lenoir, G.M., O'Conor, G.T., and Olweny, C.L.M. (eds.) (1985): *Burkitt's Lymphoma: A Human Cancer Model.* IARC Scientific Publications No. 60. IARC, Lyons.
77. Lenoir, G.M., Land, H., Parada, L.F., Cunningham, J.M., and Weinberg, R.A. (1984): Activated oncogenes in Burkitt's lymphoma. *Curr. Top. Microbiol. Immunol.,* 113:6–14.
78. Lenoir, G.M., Philip, T., and Sohier, R. (1984): Burkitt-type lymphoma -EBV association and cytogenetic markers in cases from various geographic locations. In: *Pathogenesis of Leukemias and Lymphomas: Environmental Influences,* edited by I.T. Magrath, G.T. O'Conor, and B. Ramot, pp. 283–295. Raven Press, New York.
79. Lenoir, G., Preud'homme, J.L., Bernheim, A., and Berger, R. (1982): Correlation between immunoglobulin light chain expression and variant translocation in Burkitt's lymphoma. *Nature,* 298:474–476.
80. Lenoir, G.M., and Taub, R. (1986): Chromosomal translocations and oncogenes in Burkitt's lymphoma. In: *Leukaemia and Lymphoma Research: Genetic Rearrangements in Leukaemia and Lymphoma, Vol. 2,* edited by J.M. Goldman and D.G. Harnden, pp. 152–172. London.
81. Lenoir, G.M., Vuillaume, M., and Bonnardel, C. (1985): The use of lymphomatous and lymphoblastoid cell lines in the study of Burkitt's lymphoma. In: *Burkitt's Lymphoma: A Human Cancer Model,* edited by G.M. Lenoir, G. O'Conor, and C.L.M. Olweny, IARC Scientific Publications No. 60, pp. 309–318. IARC, Lyons.
81a. Lindahl, T., Adams, A., Bjursell, G., Bornkamm, G.W., Kaschka-Dierich, C., and Jehn, YU. (1976): Covalently closed circular duplex DNA of Epstein-Barr virus in a human lymphoid cell line. *J. Mol. Biol.,* 102:511–530.
82. Magrath, I. (1983): Burkitt's Lymphoma: Clinical Aspects and Treatment. In: *Diseases of the Lymphatic System,* Diagnosis and Therapy, edited by D.W. Molander, pp. 103–139. Springer-Verlag, New York.
83. Maguire, R.T., Robins, T.S., Thorgeirsson, S.S., and Heilan, C.A. (1983): Expression of cellular *myc* and *ras* genes in undifferentiated B-cell lymphomas of Burkitt and non-Burkitt types. *Proc. Natl. Acad. Sci. U.S.A.,* 80:1947–1950.
84. Mann, K., Staunton, D., and Thorley-Lawson, D. (1985): Epstein-Barr virus-encoded protein found in plasma membranes of transformed cells. *J. Virol.,* 55:710–720.
85. Manolov, G., and Manolova, Y. (1972): Marker band in one chromosome 14 from Burkitt lymphomas. *Nature,* 237:33–34.
86. Manolova, Y., Manolov, G., Kieler, J., Levan, A., and Klein, G. (1979): Genesis of the 14q+ marker in Burkitt's lymphoma. *Hereditas,* 90:5–10.
87. Matsuo, T., Heller, M., Petti, L., O'Shiro, E., and Kieff, E. (1984): Persistence of the entire Epstein-Barr virus genome integrated into human lymphocyte DNA. *Science,* 226:1322–1324.
88. Melchers, F., and Andersson, J. (1984): B-Cell activation: Three steps and their variations. *Cell,* 37:715–720.
88a. Misko, J.S., Moss, D.J., and Pope, J.H. (1980): HLA antigen-related restriction of T lymphocyte cytotoxicity to Epstein-Barr virus. *Proc. Natl. Acad. Sci. USA,* 77:4247–4250.
89. Mitelman, F. (1981): Marker chromosome 14q+ in human cancer and leukemia. *Adv. Cancer Res.,* 34:141–170.
90. Miyoshi, I., Hiraki, S., Kimura, I., Miyamoto, K., and Sato, J. (1979): 2/8 translocation in a Japanese Burkitt's lymphoma. *Experientia,* 35:742–743.
91. Morrow, R.H., Jr. (1985): Epidemiological evidence for the role of falciparum malaria in the pathogenesis of Burkitt's lymphoma. In: *Burkitt's Lymphoma: A Human Cancer Model,* edited by G.M. Lenoir, G. O'Conor, and C.L.M. Olweny, IARC Scientific Publications No. 60, pp. 177–186. IARC, Lyons.

91a. Moss, D.J., Burrows, S.R., Castelino, D.J., Kane, R.G., Pope, J.H., Rickinson, A.B., Alpers, M.P., and Heywood, P.F. (1983): A comparison of Epstein-Barr Virus-specific T-cell immunity in malaria-endemic and nonendemic regions of Papua New Guinea. *Int. J. Cancer*, 31:727–732.
92. Mougneau, E., Lemieux, L., Rassoulzadegan, M., and Cuzin, F. (1984): Biological activities of v-*myc* and rearranged c-*myc* oncogenes in rat fibroblast cells in culture. *Proc. Natl. Acad. Sci. U.S.A.*, 81:5758–5762.
93. Mueller-Lantzsch, N., Lenoir, G., Sauter, M., Takaki, K., Béchet, J.M., Kuklik-Roos, C., Wunderlich, D., and Bornkamm, G. (1985): Identification of the coding region for a second Epstein-Barr virus nuclear antigen (EBNA2) by transfection of cloned DNA fragments. *EMBO J.*, 4:1805–1811.
94. Müller, R., Bravo, R., Burckhardt, J., and Curran, T. (1984): Induction of c-*fos* gene and protein by growth factors precedes activation of c-*myc*. *Nature*, 312:716–720.
95. Murray, M., Cunningham, J., Parada, L., Dautry, F., Leibowitz, P., and Weinberg, R.A. (1983): The HL-60 transforming sequence: a *ras* oncogene coexisting with altered *myc* genes in hematopoietic tumors. *Cell*, 33:749–757.
96. Neckers, L.M., and Cossman, J. (1983): Transferrin receptor induction in mitogen-stimulated human T lymphocytes is required for DNA synthesis and cell division, and is regulated by interleukin 2. *Proc. Natl. Acad. Sci. U.S.A.*, 80:3494–3498.
97. O'Conor, G.T. (1961): Malignant lymphoma in African children. II. A pathological entity. *Cancer*, 14:270–283.
98. O'Conor, G.T. (1970): Persistent immunologic stimulation as a factor in oncogenesis with special reference to Burkitt's tumor. *Am. J. Med.*, 48:279–285.
99. O'Conor, G.T., Rappaport, H., and Smith, E.B. (1985): Childhood lymphoma resembling 'Burkitt tumor' in the United States. *Cancer*, 18:411–417.
100. Ohno, S., Babonits, M., Wiener, F., Spira, J., Klein, G., and Potter, M. (1979): Nonrandom chromosome changes involving the Ig gene-carrying chromosome 12 and 6 in pristane-induced mouse plasmacytomas. *Cell*, 18:1001–1007.
101. Ohno, S., Migita, S., Wiener, F., Babonits, M., Klein, G., Mushinski, J.F., and Potter, M. (1984): Chromosomal translocations activating *myc* sequences and transduction of v-*abl* are critical events in the rapid induction of plasmacytomas by pristane and Abelsons virus. *J. Exp. Med.*, 159:1762–1777.
102. Opstelten, D., and Osmond, D.G., (1983): Pre-B cells in mouse bone marrow: immunofluorescence stathmokinetic studies of the proliferation of cytoplasmic μ-chain-bearing cells in normal mice. *J. Immunol.*, 131:2635–2640.
103. Palmieri, S., Kahn, P., and Graf, T. (1983): Quail embryo fibroblasts transformed by four v-*myc*-containing virus isolates show enhanced proliferation but are not tumorigenic. *EMBO J.*, 2:2385–2389.
104. Persson, H., Gray, H.E., and Godeau, F. (1985): Growth-dependent synthesis of c-*myc* encoded proteins: early stimulation by serum factors in synchronized mouse 3T3 cells. *Mol. Cell. Biol.*, 5:2903–2912.
105. Philip, T., Lenoir, G.M., Bryon, P.A., Gerard-Marchant, R., Souillet, G., Philippe, N., Freycon, F., and Brunat-Mentigny, M. (1982): Burkitt-type lymphoma in France among non-Hodgkin malignant lymphomas in Caucasian children. *Br. J. Cancer*, 45:670–678.
106. Piechaczyk, M., Yang, J.Q., Blanchard, J.M., Jeanteur, P., and Marcu, K.B. (1985): Posttranscriptional mechanisms are responsible for accumulation of truncated c-*myc* RNAs in murine plasma cell tumors. *Cell*, 42:589–597.
107. Potter, M., Wax, J.S., Anderson, A.O., and Nordan, R.P. (1985): Inhibition of plasmacytoma development in BALB/c mice by indomethacin. *J. Exp. Med.*, 161:996–1012.
108. Potter, M., Wiener, F., and Muschinski, J.F. (1984): Recent developments in plasmacytomagenesis in mice. In: *Advances In Viral Oncology*, edited by G. Klein, pp. 139–162. Raven Press, New York.
109. Preud'homme, J.L., Dellagi, K., Guglielmi, P., Vogler, L.B., Danon, F., Lenoir, G.M., Valensi, F., and Brouet, J.C. (1985): Immunologic markers of Burkitt's lymphoma cells. In: *Burkitt's Lymphoma: A Human Cancer Model*, edited by G.M. Lenoir, G. O'Conor, and C.L.M. Olweny, IARC Scientific Publication No. 60, pp. 47–64. IARC, Lyons.
110. Purtilo, D.T. (1981): Malignant lymphoproliferative diseases induced by Epstein-Barr virus in immunodeficient patients, including X-linked, cytogenetic, and familial syndromes. *Cancer Genet. Cytogenet.*, 4:251–268.

111. Rabbitts, T.H., Forster, A., Hamlyn, P., and Baer, R. (1984): Effect of somatic mutation within translocated c-*myc* genes in Burkitt's lymphoma. *Nature,* 309:592–597.
112. Rabbitts, P.H., Forster, A., Stinson, M.A., and Rabbitts, T.H. (1985a): Truncation of exon 1 from the c-*myc* gene results in prolonged c-*myc* m-RNA stability. *EMBO J.,* 4:3727–3733.
113. Rabbitts, T.H., Hamlyn, P.H., and Baer, R. (1983): Altered nucleotide sequences of a translocated c-*myc* gene in Burkitt lymphoma. *Nature,* 306:760–765.
114. Rabbitts, P.H., Watson, J.V., Lamond, A., Forster, A., Stinson, M.A., Evan, G.I., Fischer, W., Atherone, E., Sheppard, R., and Rabbitts, T.H. (1985): Metabolism of c-*myc* gene products: c-*myc* mRNA and protein expression in the cell cycle. *EMBO J.,* 4, 8:2009–2015.
115. Radbruch, A., Burger, C., Klein, S., and Mueller, W. (1986): Control of immunoglobulin class switch recombination. In: *Immunological Reviews: Control of Immunoglobulin Gene Expression, Vol. 89,* edited by G. Moeller, pp. 69–83. Stockholm.
116. Ramsay, G., Evan, G.I., and Bishop, J.M. (1984): The protein encoded by the human protooncogene c-*myc. Proc. Natl. Acad. Sci. U.S.A.,* 81:7742–7746.
117. Rapp, U.R., Cleveland, J.L., Brightman, K., Scott, A., and Ihle, J.N. (1985): Abrogation of IL-3 and IL-2 dependence by recombinant murine retroviruses expressing v-*myc* oncogenes. *Nature,* 317:434–438.
118. Rapp, U.R., Cleveland, J.L., Fredrickson, T.N., Holmes, K.L., Morse III, H.C., Jansen, H.W., Patschinsky, T., and Bister, K. (1985): Rapid induction of hemopoietic neoplasms in newborn mice by a *raf*(mil)/*myc* recombinant murine retrovirus. *J. Virol.,* 55:23–33.
118a.Rawlins, D.R., Milman, G., Hayward, S.D., and Hayward, G.S. (1985): Sequence-Specific DNA binding of the Epstein-Barr Virus Nuclear Antigen (EBNA-1) to clustered sites in the plasmid maintenance region. *Cell,* 42:859–868.
119. Reed, J.C., Nowell, P.C., and Hoover, R.G. (1985): Regulation of c-*myc* mRNA levels in normal human lymphocytes by modulators of cellular proliferation. *Proc. Natl. Acad. Sci. U.S.A.,* 84:4221–4224.
120. Reed, J.C., Sabath, D.E., Hoover, R.G., and Prystowsky, M.B. (1985): Recombinant interleukin 2 regulates levels of c-*myc* mRNA in a cloned murine T-lymphocyte. *Mol. Cell. Biol.,* 5:3361–3368.
121. Rickinson, A.B., Wallace, L.E., and Epstein, M.A. (1980): HLA-restricted T-cell recognition of Epstein-Barr virus infected B cells. *Nature,* 283:865–867.
122. Rickinson, A.B., Yao, Q.Y., and Wallace, L.E. (1985): The Epstein-Barr virus as a model of virus-host interactions. *Br. Med. Bull.,* 41:75–79.
123. Roberts, A.B., Anzano, M.A., Wakefield, L.M., Roche, N.S., Stern, D.F., and Sporn, M.B. (1985): Type B transforming growth factor: A bifunctional regulator of cellular growth. *Proc. Natl. Acad. Sci. U.S.A.,* 82:119–123.
124. Rocchi, G., de Felici, A., Ragona, G., and Heinz, A. (1977): Quantitative evaluation of Epstein-Barr virus-infected mononuclear peripheral blood leukocytes in infectious mononucleosis. *N. Engl. J. Med.,* 296:132–134.
125. Rooney, C.M., Rickinson, A.B., Moss, D.J., Lenoir, G.M., and Epstein, M.A. (1984): Paired Epstein-Barr virus-carrying lymphoma and lymphoblastoid cell lines from Burkitt's lymphoma patients: Comparative sensitivity to non-specific and to allo-specific cytotoxic responses in vitro. *Int. J. Cancer,* 34:339–349.
126. Rooney, C.M., Rowe, M., Wallace, L.E., and Rickinson, A.B. (1985): Epstein-Barr virus-positive Burkitt's lymphoma cells not recognized by virus-specific T-cell surveillance. *Nature,* 317:629–631.
127. Rosenberg, S.A. (Chairman) (1982): National Cancer Institute sponsored study of classifications of non-Hodgkin's lymphomas. Summary and description of a working formulation for clinical usage. *Cancer,* 49:2112–2135.
128. Rosenberg, Y.J. (1978): Autoimmune and polyclonal B-cell responses during murine malaria. *Nature,* 274:170–172.
129. Royer-Pokora, B., Beug, H., Claviez, M., Winkhardt, H.J., Frijs, R.R., and Graf, T. (1978): Transformation parameters in chicken fibroblasts transformed by AEV and MC29 avian leukemia viruses. *Cell,* 13:751–760.
130. Rymo, L., Klein, G., and Ricksten, A. (1985): Expression of a second Epstein-Barr virus-determined nuclear antigen in mouse cells after gene transfer with a cloned fragment of the viral genome. *Proc. Natl. Acad. Sci. U.S.A.,* 82:3435–3439.
131. Shade, M., Woodward, M.A., and Steel, C.M. (1980): Chromosome aberrations acquired in vitro by human B-cell lines. II: Distribution of break points. *J. Natl. Cancer Inst.,* 65:101–109.

132. Sigaux, F., Berger, R., Bernheim, A., Valensi, F., Daniel, M.T., and Flandrin, G. (1984): Malignant lymphomas with band 8q24 chromosome abnormality: a morphologic continuum extending from Burkitt's to immunoblastic lymphoma. *Br. J. Haematol.*, 57:393–405.
133. Sixbey, J.W., Nedrud, J.G., Raab-Traub, N., Hanes, R.A., and Pagano, J.S. (1984): Epstein-Barr virus replication in oropharyngeal epithelial cells. *N. Engl. J. Med.*, 310:1225–1230.
133a. Skare, J., Farley, J., Strominger, J.L., Fresen, K.O., Cho, M.S., and zur Hausen, H. (1985): Transformation by Epstein-Barr virus requires DNA sequences in the region BamHI fragments Y and H. *J. Virology* 55:286–297.
134. Speck, S., and Strominger, J. (1985): Analysis of the transcript encoding the latent Epstein-Barr virus nuclear antigen. I: A potentially polycistronic message generated by long-range splicing of several exons. *Proc. Natl. Acad. Sci. U.S.A.*, 82:8305–8309.
135. Sporn, M.B., and Roberts, A.B. (1985): Autocrine growth factors and cancer. *Nature*, 313:745–747.
136. Stanton, L.W., Farhlander, P., Tesser, P., and Marcu, K.B. (1984): Nucleotide sequence comparison of normal and translocated murine c-*myc* gene. *Nature*, 310:423–425.
136a. Starzl, T.E., Porter, K.A., Iwatsuki, S., Rosenthal, J.T., Shaw Jr., B.W., Atchison, R.W., Nalesnik, M.A., Ho, M., Griffith, B.P., Hakala, T.R., Hardesty, R.L., Jaffe, R., and Bahnson, H.T. (1984): Reversibility of lymphomas and lymphoproliferative lesions developing under cyclosporinsterioid therapy. *Lancet*, I:583–587.
137. Stewart, T.A., Pattengale, P.K., and Leder, P. (1984): Spontaneous mammary adenocarcinomas in transgenic mice that carry and express MTV/*myc* fusion genes. *Cell*, 38:627–637.
138. Stiles, C.D., Capone, G.T., Scher, C.D., Antoniades, H.N., van Wyk, J.J., and Pledger, W.J. (1979): Dual control of cell growth by somatomedins and platelet-derived growth factor. *Proc. Natl. Adad. Sci. U.S.A.*, 76:1279–1283.
139. Summers, W.P., Grogan, E.A., Shedd, D., Robert, M., Chun-Ren, L., and Miller, G. (1982): Stable expression in mouse cells of nuclear neoantigen after transfer of a 3,4-megadalton cloned fragment of Epstein-Barr virus DNA. *Proc. Natl. Acad. Sci. U.S.A.*, 79:5688–5692.
140. Taub, R., Kirsch, I., Morton, C., Lenoir, G., Swan, D., Tronick, S., Aaronson, S., and Leder, P. (1982): Translocation of the c-*myc* gene into the immunoglobulin heavy chain locus in human Burkitt's lymphoma and murine plasmacytoma cells. *Proc. Natl. Acad. Sci. U.S.A.*, 79:7837–7841.
141. Taub, R., Moulding, C., Battey, J., Murphy, W., Vasicek, T., Lenoir, G.M., and Leder, P. (1984): Activation and somatic mutation of the translocated c-*myc* gene in Burkitt lymphoma cells. *Cell*, 36:339–348.
142. Thompson, C.B., Challoner, P.B., Neiman, P.E., and Groudine, M. (1985): Levels of c-*myc* oncogene mRNA are invariant through the cell cycle. *Nature*, 314:363–366.
143. Thorley-Lawson, D.A., and Mann, K.P. (1985): Early events in Epstein-Barr virus infection provide a model for B-cell activation. *J. Exp. Med.*, 162:45–59.
144. Tonegawa, S. (1983): Somatic generation of antibody diversity. *Nature*, 302:575–581.
145. Van den Berghe, H., Parloir, A., Gosseye, S., Englebienne, V., Cornu, G., and Sokal, G. (1979): Variant translocation in Burkitt lymphoma. *Cancer Genet. Cytogenet.*, 1:9–14.
146. Wabl, M.R., and Burrows, P.D. (1984): Expression of immunoglobulin heavy chain at a high level in the absence of a proposed immunoglobulin enhancer element in *cis*. *Proc. Natl. Acad. Sci. U.S.A.*, 81:2452–2455.
146a. Whang-Peng, J., Lee. E.C., Sieverts, H., and Magrath, I.T. (1984): Burkitt's lymphoma in AIDS: cytogenetic study. *Blood*, 63:818–822.
147. Whittle, H.C., Brown, J. Marsh, K., Greenwood, B.M., Seidelin, P., Tighe, H., and Wedderburn, L. (1984): T-cell control of Epstein-Barr virus-infected B-cells is lost during P.falciparum malaria. *Nature*, 312:449–450.
148. Wiman, K.G., Clarkson, B., Hayday, A.C., Saito, H., Tonegawa, S., and Hayward, W.S. (1984): Activation of a translocated c-*myc* gene: Role of structural alterations in the upstream region. *Proc. Natl. Acad. Sci. U.S.A.*, 81:6798–6802.
149. Wright, D.H. (1963): Cytology and histochemistry of the Burkitt's lymphoma. *Br. J. Cancer*, 17:50–55.
150. Yang, J.Q., Bauer, S., Mushinski, J.F., and Marcu, K.B. (1985): Chromosome translocations clustered 5' of the murine c-*myc* qualitatively affect promoter usage: implications for the site of normal c-*myc* regulation. *EMBO J.*, 4:1441–1447.
151. Yao, Q.Y., Rickinson, A.B., Gaston, J.S.H., and Epstein, M.A. (1985): *In vitro* analysis of the Epstein-Barr virus-host balance in long-term renal allograft recipients. *Int. J. Cancer*, 35:43–49.

152. Yao, Q.Y., Rickinson, A.B., and Epstein, M.A. (1985): A reexamination of the Epstein-Barr virus carrier state in healthy seropositive individuals. *Int. J. Cancer,* 35:35–42.
153. Yates, J., Warren, N., Reisman, D., and Sudgen, B. (1984): A *cis*-acting element from the Epstein-Barr viral genome that permits stable replication of recombinant plasmids in latently infected cells. *Proc. Natl. Acad. Sci. U.S.A.,* 81:3806–3810.
154. Yates, J., Warren, N., and Sudgen, B. (1985): Stable replication of plasmids derived from Epstein-Barr virus in various mammalian cells. *Nature,* 313:812–815.
155. Zaller, D.M., and Eckhardt, L.A. (1985): Deletion of B-cell-specific enhancer affects transfected, but not endogenous immunoglobulin heavy-chain gene expression. *Proc. Natl. Acad. Sci. U.S.A.,* 82:5088–5092.
156. Zech, L., Haglund, U., Nilsson, K., and Klein, G. (1976): Characteristic chromosomal abnormalities in biopsies and lymphoid cell lines from patients with Burkitt and non-Burkitt lymphomas. *Int. J. Cancer,* 17:47–56.
157. Ziegler, J.L., Beckstead, J.A., Volberding, P.A., Abrams, D.I., Levine, A.M., Lukes, R.J., Gill, P.S., Burkes, R.L., Meyer, P.R., Metroka, C.E., Mouradian, J., Moore, A., Riggs, S.A., Butler, J.J., Cabanillas, F.C., Hersh, E., Newell, G.R., Laubenstein, L.J., Knowles, D., Odajnyk, C., Raphael, B., Koziner, B., Urmacher, C., and Clarkson, B. (1984): Non-Hodgkin's lymphoma in 90 homosexual men. *N. Engl. J. Med.,* 311:565–570.
158. Ziegler, J.L., Miner, R.C., Rosenbaum, E., Lennette, E.T., Shillitoe, E., Casavant, C., Drew, W.L., Mintz, L., Gershow, J., Greenspan, J., Beckstead, J., and Yamamoto, K. (1982): Outbreak of Burkitt's Lymphoma in homosexual men. *Lancet,* 2:631–633.
159. zur Hausen, H., Schulte-Holthausen, H., Klein, G., Henle, W., Henle, G., Clifford, P., and Santesson, L. (1970): EBV DNA in biopsies of Burkitt tumours and anaplastic carcinomas of the nasopharynx. *Nature,* 228:1056–1058.
160. zur Hausen, H. (1986): Human papillomaviruses: Why are some types carcinogenic? In: *Concepts in Viral Pathogenesis,* edited by M. Oldstone and A. Norkins. Springer-Verlag, New York *(in press).*

In Defense of the "Old" Burkitt Lymphoma Scenario

George Klein

Department of Tumor Biology, Karolinska Institutet, S-104 01 Stockholm, Sweden

In the preceding article, Gilbert Lenoir and Georg Bornkamm present a new scenario for the multistep development of Burkitt lymphoma. In my double capacity of Editor of this volume and the person responsible for the previous scenario, I was delighted to see that Lenoir and Bornkamm did exactly what I hoped they would do: present an alternative that is radically different from my own. They have done so with profundity and elegance and I also had the opportunity to discuss the alternative scenario with them at length. While I am impressed by much of their argumentation and will certainly concede that many of their points ring true, I still feel that much of the evidence continues to favor the originally proposed order where EBV-infection is followed by the translocation, rather than the reverse. I shall briefly summarize my criticism of the new scenario and present some additional points in defense of the old.

CRITICISM OF THE LENOIR (LBsc) BORNKAMM SCENARIO

1. EBV-infection is a very common event. Most African children are infected at an early age, and the pre-BL children have particularly high anti-EBV titers, indicating that they carry a high virus load, well in advance of the disease (3).
2. Infected persons have high titers of neutralizing antibodies already some weeks after the primary infection. This should block the transfer of the virus between cells, although it does not necessarily prevent intracellular cross infection by direct contact.
3. Permanent conversion of established EBV-negative BL lines into EBV-carrying sublines by *in vitro* infection is possible, but difficult and usually requires repeated reinfection and prolonged selection.
4. The majority of all EBV-negative BLs were found to arise in EBV seropositive individuals (8). If secondary infection of a translocation carrying clone with EBV provides the cells with a strong growth advantage, as postulated by the LBsc, it is hard to understand why 80% of the nonendemic BLs have remained EBV-negative (2). It is even more surprising that W. Henle and

G. Henle (personal communication) have seen several initially sero-negative American BL patients who acquired primary EBV infections in the course of time, with or without signs of mononucleosis, but their tumors remained free of EBV. In these cases, the translocations carrying cells were not infected even in the absence of antibodies to EBV.
5. According to the LBsc, the translocation must be followed by EBV conversion fairly soon in order to provide the cell with the final push to autonomous growth and to prevent its elimination. This is perhaps the greatest weakness of the LBsc, because it requires the occurrence of two low-probability events in the same clone, within a fairly short time interval.
6. The LBsc is compelled to postulate the existence of an unknown mechanism, to explain the origin EBV-negative BLs. The EBV-carrying and the EBV-negative disease are closely similar, however. Minor clinical and pathological differences can be explained by the stronger immunogenicity of the virus-carrying cells *in vivo*. Both forms of the disease carry identical translocations and arise from the same cell type. The dualistic explanation postulated by the LBsc therefore rather unsatisfactory.
7. Conversion of EBV-negative BL-lines *in vitro* frequently induces the appearance of B-cell activation markers that are not found on freshly examined EBV-negative or EBV-carrying BL biopsies or on derived lines of recent origin (4,10,11). IF EBV conversion follows the translocation *in vivo*, as postulated by the LBsc, one would expect a similar difference between EBV negative and EBV positive BL cells *in vivo*, with regard to activation markers, instead of the absence of such markers and the phenotypic indistinguishability of the two forms.

DEFENSE OF THE OLD SCENARIO

I see the main BL-promoting role of EBV in its ability to expand the target cell population where the critically important translocation can occur as a rare genetic accident. EBV-transformed cells are more long lived and somewhat less prone to differentiate than normal B-cells. Similarly to all other genetic accidents, the probability of the BL translocations can be expected to increase with the number of cell divisions.

Like Lenoir and Bornkamm, I view the accelerating effect of the Abelson leukemia virus (AbLV) on mouse plasmacytoma (MPC) development as potentially akin to the EBV-BL scenario, but interpret it somewhat differently. It is particularly noteworthy that AbLV immortalizes pre-B, not B-cells. Nevertheless, the tumors that appear in AbLV + pristane oil treated mice are equally mature plasmacytomas and carry the same translocations as the usual pristane oil induced MPCs that arise later and in a lower frequency. Expansion of the pre-B precursor population is the most likely acceleration mechanism, particularly since it is known that AbLV transformed pre-B cells continue to rearrange their Ig-genes. The translocation would occur later, in the chronically stimulated mature B-cell population. This scenario is also in line with some recent findings concerning the effect of constitutively activated c-myc genes, introduced into the tumorigenic precursor cell population at various times. In the experiments of Potter et al, infection of pristane oil treated Balb/c mice with the v-myc containing J3 retroviral constructs has lead the development of typical MPCs that had no translocations (9). The constitutive expression of the retrovirally activated myc gene has apparently obviated the need for the activation of the gene by juxtaposition to the Ig-locus. These tumors appeared earlier and were somewhat more immature (plasmoblasts) than the usual translocation-carrying MPCs. A different situation was encountered by Adams et al (1) in transgenic mice. Here, the c-myc gene was linked to an immunoglobulin enhancer and introduced into mouse eggs. The transgenic mice developed largely pre-B and also some B-lymphomas, but no plasmacytomas. In this case, the constitutive activation of c-myc is expected to occur when cells commit themselves to the B-lymphocyte lineage. This clearly favors the development of pre-B tumors.

These findings can be discussed in relation to the postulate of the LBsc, suggesting that the translocation occurs during the normal DNA rearrangement i.e. at an early stage of B-cell differentiation. Burkitt lymphomas are B-cells, not pre-B cells. The lack of the latter can not be attributed to the insensitivity of pre-B cells to EBV transformation, since Katamine et al. (5), have shown that EBV can immortalize cells committed to the B-cell lineage even prior to IgH-rearrangement.

Occurrence of the translocations at the time of the normal IgH rearrangement is also unlikely for other reasons. If it is assumed that the rearrangement creates translocation-prone hot spots, one would expect a preferential involvement of VDJ sequences, rather than of switch sites. The opposite is true, however. The predominance of the switch sites suggest that switch recombination enzymes may participate in the translocations. Class switches occur in mature B-cells, however, and often only after prolonged antigen stimulation.

While it is thus difficult to explain the preferential involvement of the switch sites on the basis of the LBsc, they are well in line with our current and more detailed interpretation of the original scenario (6,7). We favor the idea that the translocations occur at the time when an antigenically stimulated, clonally expanded B-cell is about to return to G_O, upon the waning of the antigenic stimulus, and change into a memory cell. This is supported by the regular presence of a heavy chain in all BL cells, suggesting preselection by an antigen, and the fact that the phenotype resembles resting, rather than activated B cells. We assume that the immediate BL precursor cells have already received a signal to return to the resting stage at or around the time of the translocation. In response, they have switched off their normal, nontranslocated c-myc allele (as shown by several laboratories, see ref. 6) but continue to express the displaced gene that has come under the control of a constitutively expressed Ig-locus. It is known that myc-expression is down-regulated in many different types of cells before the cell enters the resting stage (for review see ref. 7). It is reasonable to assume that interference with this down regulation keeps the cell in cycle.

The idea of the "translocation-suspended B memory cell" is also in line with the fact that most mouse plasmacytomas make IgA while most rat immunocytomas produce IgE. This would correspond to the immunoglobulin class of the predominant memory cells in the precursor populations of these two tumors. BL cells make predominantly IgM and Smu is the most common myc-acceptor site. This does not contradict the hypothesis, since it has been shown that IgG memory is often stored in IgM positive cells (12,13). One might even speculate that the readiness of IgM positive memory cell to perform the Smu/Sgamma recombination switch at the time of antigenic reactivation may increase the translocation risk in the Smu region.

CONCLUSION

On the basis of the above arguments I still prefer the old scenario but I will readily admit that the evidence is insufficient to decide one way or another. Studies in the following areas will be particularly relevant for further progress in this area: (a) improved definition of EBV latency *in vivo;* (b) more detailed analysis of the B-cell life cycle, particularly the transition from the proliferating state to G_O; (c) understanding of the interaction between constitutively activated myc genes and the differentiation program of B-cells; (d) improved understanding of the relationships between B-cell phenotype, differentiation and tumorigenicity.

ACKNOWLEDGMENT

In my double capacity of author and editor, I would like to express my sincere thanks to Lenoir and Bornkamm for having provided a most stimulating alternative to my point of view.

REFERENCES

1. Adams, J.M., Harris, A.W., Pinkert, C.A., Corcoran, L.M., Alexander, W.S., Cory, S., Palmiter, R.D., and Brinster, R.L. (1985): The c-myc oncogene drive by immunoglobulin enhancers induces lymphoid malignancy in transgenic mice. *Nature* 318:533–538.
2. Andersson, M., Klein, G., Ziegler, J.L., and Henle, W. (1976): Association of Epstein-Barr viral genomes with American Burkitt lymphoma. *Nature* 260:357–359.
3. de Thé G (1980): Role of Epstein-Barr virus in human disease: infectious mononucleosis, Burkitt's lymphoma, and nasopharyngeal carcinoma. In: *Viral Oncology,* edited by G. Klein pp. 769–797. Raven Press, New York.
4. Ehlin-Henriksson B., Manneborg-Sandlund A., and Klein G. (1986): Expression of B-cell specific markers in different Burkitt lymphoma subgroups. (Submitted for publication.)
5. Katamine, S., Otsu, M., Tada, K., Tsuchiya, S., Sato, T., Ishida, N., Honjo, T., and Ono, Y. (1984): Epstein-Barr virus transforms precursor B cells even before immunoglobulin gene rearrangement. *Nature* 309:369–372.
6. Klein, G., and Klein, E. (1985): Myc/Ig juxtaposition by chromosomal translocations: some new insights, puzzles and paradoxes. *Immunol. Today* 6:208–215.
7. Klein, G., and Klein, E. (1986): Conditioned tumorigenicity of activated oncogenes. *Cancer Res.,* 46:3211–3224.

8. Levine, P.H., Kamaraju, L.S., Connelly, R.F., Berard, C.W., Dorfman, R.F., Magrath, I., and Easton, J.M. (1982): The American Burkitt's lymphoma registry: Eight years' experience. *Cancer* 49:1016–1019.
9. Potter, M., Mushinski, J.F., Mushinski, E.B., Brust, S., Wax, J.S., Wiener, F., Babonits, M., Rapp, U.R., and Morse H.C. (1986): Rapid induction of plasmacytomas in mice by pristane and murine recominant retrovirus J-3 containing an avian c-myc oncogene. (Submitted for publication.)
10. Rowe, M., Rooney, C., Rickinson, A., Lenoir, G., Rupani, H., Moss, D., Stein, H., and Epstein, M. (1985): Distinctions between endemic and sporadic form of Epstein-Barr virus-positive Burkitt's lymphoma. *Int. J. Cancer* 35:435–442.
11. Rowe, M., Rooney, C., Edwards, C., Lenoir, G., and Rickinson, J. (1986): Epstein-Barr virus status and tumor cell phenotype in sporadic Burkitt's lymphoma. *Int. J. Cancer* 37:367–373.
12. Yefenof, E., Sanders, C.M., Snow, E.C., Noelly, R.J., Oliver, K.G., Uhr, G.W., and Vitetta, E.S. (1985): Preparation and analysis of antigen specific memory B-cells. *J. Immunol.* 135-3777–3784.
13. Yefenof, E., Sanders, C.M., Uhr, G.W., and Vitetta, E.S.: In vitro activation of murine antigen specific memory B-cells by a T dependent antigen. *J. Immunol.* (in press).

Subject Index

A
(9; 22) translocation, cytogenetic morphology, 55–56
13q14.11, 21
Abelson murine leukemia virus, 79
Acute lymphocytic leukemia, 45–47
AIDS
  lymphoma, 190
    Burkitt-type, 190
Allele
  Rb-1, 20–21
  WAGR, 27
Alpha chain locus, 43–47
Amplified gene, 146–147
Aneuploidy, 141
Antioncogene, 11–12

B
BALB/c mice, plasmacytoma development, 99–118
Band, 34
B-cell
  growth, *myc* expression, 192
  neoplasia
    molecular basis, 35–49
    non-Burkitt's lymphoma, genetic analysis, 41–47
  proliferation, control, 187
bcr, 79
  gene part, 84–86
  restriction map, 83
bcr breakpoint
  K562, 86–91
  Ph^1, 85
bcr/c-abl mRNA, chimeric, 89–91
Bloom's syndrome, 2
B lymphocyte, 108
  Epstein-Barr virus, 173, 186
B lymphocyte chromosome, induced chromosome breaks, 101
Breakpoint, 34–35
  identical, 55–56
Breakpoint cluster region, 79
Burkitt's lymphoma, 35–40, 174–177
  African, 190–191
  carrier state definition, 187
  cell
    chromosomal translocation, 176–177
    Epstein-Barr virus genome-containing, 173
    tissue culture of, 175–176
    translocation molecular consequences, 177–185
  c-*myc* locus, 178–179
  c-*myc* oncogene, 36–40
  epidemiology, 190–191
  Epstein-Barr virus association inconsistency, 175
  genetic pathogenesis mechanisms, 40
  incidence, geographic variation, 174–175
  molecular studies, 191–192
  multistep scenario, 173–198
  new scenario criticism, 207–208
  new scenario proposal, 173–198
  old scenario defense, 207–210
  pathoclinical definition, 174
  pathogenesis, 185
    cellular oncogene systems, 185
  scenarios compared, 193

C
c-*abl*, 79–82, 145
  activation, 77–95
  chromosome 9, 78–79
  chromosome 9 breakpoint, 80–82
  chronic myelocytic leukemia, 77–95
  restriction map, 81
Cancer
  age-specific incidence, 1–2
  first mutational event, 4–7
  growth, 7–8
  hereditary, 2–3
    first mutational event, 4–5
    recessive, 2
  human, two-mutation model, 1–13
  intermediate cell, 5–6
  mathematical model, 8–9

Cancer (cont.)
  mutation, 1
  nonhereditary, first mutational event, 5
  progression, 7–8
  second, genetic event, 6–7
Cancer gene
  development, 12–13
  mutation, 10–13
  oncogenesis, 12–13
Cancer susceptibility gene, recessive, 19–31
Carcinogenesis, 10–13
Carcinoma, nasopharyngeal, 185–186
Cellular gene, mouse mammary tumor virus activation, 130–133
Cellular proliferation, *myc* gene, 179–181
c-*erb*B, 145
c-*erb*B-2, 133–134
c-Ha-*ras*1, 133–134
Child
  Burkitt's lymphoma, 174–175
  cancer, 174–175
Chimeric *bcr/c-abl* mRNA, 89–91
Chimeric mRNA translation, 91
Chromosome
  double-minute, 141, 146–147
  homogeneously staining region, 141, 146–147
  translocation, Burkitt's lymphoma, 176–177
Chromosome 9, 53, 57
  breakpoint, 80–82
  c-*abl*, 78–79
Chromosome 11, 44
Chromosome 13, 23
  loss, 22, 24
Chromosome 14, 44, 45
  band 14q32, 41
  breakpoint, 43–47
Chromosome 15, D2/D3 band, 99
Chromosome 22, 53, 57
  breakpoint cluster region, 79, 82–86
Chromosome aberration, chronic phase, 72
Chronic lymphocytic leukemia cell, 41
CLL 271, 44
CLL 1386, 44
Clone, 55
CML
  blast-phase, 72–73
  breakpoint, DNA sequencing analysis, 88–89
  c-*abl*, 77–79

chimeric mRNA, 91–93
chromosomal aberrations, functional perspective, 69–73
chromosome 22, 79
classical Ph^1-positive
  secondary changes, 56–62
  secondary chromosome aberrations, 61, 66–67
classical t(9;22) associated
  combination frequency, 60
  secondary chromosome aberrations, 58–59
metaphase, 56
multistep cytogenetics, 53–73
oncogene, 72
pathogenesis, 71–72
Ph^1, 53
Ph^1-negative, chromosome aberrations, 67–69
Ph^1-positive, variant translocations, 63–67
Ph^1 translocation, c-*abl* activation, 77–95
translocation, 53
translocation-oncogene rearrangement, 69–71
c-*myb*, 144
c-*myc*, 133–134, 142–143
  lung cancer, 155
c-*myc* gene
  amplification, 159
  expression, 159
  translocated, immunoglobin locus regulation, 182
  translocation breakpoint, 178–179
c-*myc* locus, Burkitt's lymphoma, 178–179
c-*myc* oncogene
  anomalies, 102–116
  Burkitt's lymphoma, 36–40
  chromosomal translocation, 102–116

E

Eco RI digestion, hybridization comparison, 160, 162, 164, 165
e-*myc*, homologous region, 169
Epithelial cell, Epstein-Barr virus, 186
Epstein-Barr virus, 173, 185, 189
  B lymphocyte, 186
  Burkitt's lymphoma association inconsistency, 175
  carrier state definition, 187

SUBJECT INDEX

dual tropism, 186
epithelial cell, 186
human B lymphocyte, 173
immune dysfunction, 187–188
infectious mononucleosis, 186
lymphoproliferative disease, 187–188
nasopharyngeal carcinoma, 185–186
Epstein-Barr virus-immortalized cell, gene transcription, 188

F

Focal proliferation, 110–111
   inhibition, 111–114

G

Gene amplification, 78
Gene rearrangement, 77–78
G-group chromosome, 53

H

Hepatoblastoma, 19
   WAGR locus, 28–29
Human B lymphocyte, Epstein-Barr virus, 173
Human c-*abl* proto-oncogene, 78–79
Human cancer susceptibility gene, recessive, 19–31
Human model, 1–13
Human N-*myc* probe Nb-1, 166
Hyperplasia
   transplantation, 125–126
   tumor development, 123–125
Hyperplastic alveolar nodule, 123–125
   clonality, 127
   hyperplastic outgrowth, 124–125

I

Infectious mononucleosis, 186
Immune dysfunction, Epstein-Barr virus, 187–188
Immunoglobin, recombinational machinery, 191–192
Immonoglobin locus regulation, 182
Immunoglobulin heavy-chain gene, 38–39
Immunoglobulin-secreting cell, normal, 107–114
Immunoglobulin secretion, 107
int-1, 130–134
int-2, 130–134

K

K562
   bcr breakpoint, 86–91
   Ph^1 chromosomal breakpoint, 86–87
K562 breakpoint region, restriction map, 83

L

L-*myc*
   homologous region, 169
   lung cancer, 155
L-*myc* gene
   amplification, 165–170
   expression, 165–170
Lung cancer, small cell
   biochemical characterization, 155–158
   cell culture, 155–158
   classic cell line properties, 158
   myc gene, 155–171
   variant cell line properties, 158
Lymphocytic leukemia
   acute. See Acute lymphocytic leukemia
   chronic. See Chronic lymphocytic leukemia
Lymphoid cell, Epstein-Barr virus-immortalized, nonmalignant, 173
Lymphoma
   AIDS, 190
   follicular, 41
Lymphoproliferative disease, Epstein-Barr virus, 187–188

M

Mammary oncogene, 131, 133
Mammary tissue, normal vs. hyperplastic morphology, 123–124
mcf2, 133–134
mcf3, 133–134
Mononucleosis, infections, 180
Mouse mammary tumor development, 123–137
Mouse mammary tumor virus integrated locus, tumor-associated, 129–133
Mouse plasmacytoma, chromosomal anomalies, 104–105
Mouse plasmacytoma rcpt (6;15), 103
Mouse plasmacytoma rcpt (12;15), 103
mRNA translation, chimeric, 91
MTV/*myc*, 133–134
Mutation, cancer, 1

Mutational event, first, 4–5
myc cell
  amplification, 155–171
  expression, 155–171
    B-cell growth, 192
myc gene
  biology, 170
  cellular proliferation, 179–181
  small cell lung cancer, 155–171
Myeloid leukemia, chronic. See Chronic myeloid leukemia

N

Neoplastic progression, 127–129
  defined, 123
  int-1, 131–133
  int-2, 131–133
  marker, 126–129
Neurofibromatosis, 2
N-*myc*, 143–144
  homologous region, 169
  lung cancer, 155
N-*myc* gene
  amplification, 159–165
  expression, 159–165
Nomenclature, cytogenic, 54–55
N-*ras*, 133–134

O

Oncogene, 11–12
  amplification, tumor progression, 141–149
Osteosarcoma, 19
  development, 25–27

P

Paraffin oil, 100–102
Ph[1]
  bcr breakpoint, 85
  chronic myeloid leukemia, 53
Ph[1] translocation, 77–95
  product, 89–91
Philadelphia chromosome. See Ph[1]
Plasma cell
  development, 110
  maturation, 110
  progression, accelerator, 116–118
  proliferative stage, 109–110
Plasmacytic focus, 112

Plasmacytoma
  autonomous state, 117–118
  development, 99–118
  genetic factors, 102
  IgA-secreting, IgC$_H$ switch target sites, 104–106
  inducing agents, 100–102
    mode of action, 100–102
  pristane, 114
Plasmacytoma cell, growth-dependent primary, 114–116
Plastic, solid, 100–102
Point mutation, 77
Poly (A)t cell line RNA, hybridization, 163
Polyposis coli, 2
Pristane, 100
  plasmacytoma, 114
Proto-oncogene, 77
  activation, 141
  amplification, 141–146
    tumor progression, 147–149
Proviral DNA, exogenous mouse mammary tumor virus, 126–129

R

ras gene family, 144
ras oncogene, 77
Rb gene, recessive nature, 22–25
Rb-1 locus
  hemizygosity mechanism, 25, 26
  homozygosity mechanism, 25, 26
  localization, 21
  osteosarcoma, 25–27
    development, 25–27
  WAGR locus, 29–31
Reciprocal t(11;14) (q13;q21) translocation, 41–42
Region, 34
Retinoblastoma, 19–31
  development, 20–21
  hereditary, 2
Retinoblastoma locus, 20–27
  historical aspects, 20
Rhabdomyosarcoma, 19
  WAGR locus, 28–29

T

t(8;14) (24;q11) chromosome translocation, 48

t(9;22) (q34;q11), 53
t(11;14) chromosome translocation, 42–43
t(11;14) (q13;q11) translocation, 47
t(14;18) chromosome translocation, 42–43
Target cell, two-event cancer model, 3–4
T-cell leukemia, 48
T-cell neoplasia, molecular basis, 35–49
T-cell receptor, alpha chain, 43–47
Transformation assay, 133–135
Transgenic mouse, 135
Translocation, 77–78
   Burkitt's lymphoma, 176–177
   chronic myeloid leukemia, 53–54
   c-*myc* gene, 182
   method, 183–185
   reciprocal, 103
Translocation generation, 191–192
Translocation t(2;9;22), complex variant, 63
Transplantation, hyperplasias, 125–126
Tumor development, hyperplasia, 123–125
Tumor progression, 7–8

defined, 123
proto-oncogenes amplication, 147–149
tx-1, 133–134

V

v-*abl*, 79
VDJ joining error, 42–43

W

WAGR locus, 19
   hepatoblastoma, 28–29
   Rb-1 locus, 29–31
   rhabdomyosarcoma, 28–29
Wilms' locus, 19–31
Wilms' tumor, 27–31
   development, 27
   gene, recessive nature, 28
   locus, 27–31

X

Xeroderma pigmentosum, 2